CW00525515

About the Author

Brian Utting was born in Manchester in 1956. He now lives in
Hampshire with his wife, Beverly, and their two cats. He has two
daughters, Victoria and Laura and a granddaughter Carys.

To my mum and dad,
without whom none of this would have happened.

Brian Utting

THE SECRET DIARY OF AN ELF INSPECTOR

(WELL, IT'S NOT THAT SECRET NOW – AND HAS NOTHING TO DO WITH ELVES)

AUSTIN MACAULEY PUBLISHERS™

LONDON • CAMBRIDGE • NEW YORK • SHARJAH

A CIP catalogue record for this title is available from the British Library.

ISBN 978-1-78693-361-4 (Paperback)
ISBN 978-1-78693-362-1 (E-Book)

www.austinmacauley.com

First Published (2017)
Austin Macauley Publishers™ Ltd.
25 Canada Square Canary Wharf
London
E14 5LQ

Acknowledgements

I would like to thank everyone who has ever helped me with anything.

Chapter One

"There's sewage spewing down the fucking lane from my arsehole – *what an appropriate turn of phrase* – of a neighbour's cesspit," a furious voice shrieked down the phone at me. It sounded like a howler monkey, on heat, with a thorn in its hand.

"Okay, if I can take a few de…" but before I could actually finish the sentence he abruptly butted in.

"It's been going on for bloody weeks now," was his, frankly, rude and somewhat garbled interjection.

"Yes, well I need to…"

"What are you bloody well going to do about it?" he interrupted again.

"As I keep trying to s…"

"You need to get your backside down here straight away matey."

Whoa, now hold on just a cotton pickin' minute.

There were two things which I could have assured him of straight away; (i) I wasn't going to get my backside or any other part of my anatomy down there straight away and; (ii) I was not his matey – no way, no how – but I didn't, it/he wasn't worth the effort.

"How can I get there if I don't know where there is?"

A good point well-made I thought. And I was particularly impressed that I'd actually managed to get a complete sentence out without another interruption.

"I bet you're sat there on your big fat arse just drinking coffee and stuffing your face with cake. Well you need to get off it and do some fuckin' work."

I wouldn't have minded, but I'd asked my colleagues that morning; 'Does my bum look big this?' and they'd assured me it didn't.

Liars!

"It's all a load of crap." At this point two things came immediately to mind; *that's verbal as well as anal by the sounds of it,* and the saying; *'many a true word spoken in jest.'*

"Please, just calm yourself down, because if you continue to be abusive I'll have no choice but to put the phone down on you. Okay?" I informed him politely.

"Don't you dare put the fucking phone down on me," was his instant and barbed reply.

"I won't if you're civil. But as I've already said, if you continue to be abusive, I will." I was still managing to be polite – just.

"I've got a jolly good mind to report you to your boss you little turd." He was obviously obsessed with all things lavatorial.

"What's his name?"

My boss or my turd?

Who names their turds anyway? Actually, don't answer that.

Suddenly I could hear the plum – more Mirabelle than Victoria – in his mouth now that he'd slowed down his petulant tirade.

"His name is Mr Patient – *wish it was Lou Rawls though* – and you can report me now if you want because he's standing right here."

Daniel Patient wasn't. Not by any stretch of the imagination. And that's even if your imagination contains elastic – not Lycra though, because that's the fabric of the devil and should only be used in female garments that are a size 8 or below. I bet he wouldn't have been a patient patient either if he'd had to go into hospital. And surprisingly, for his age, he was incredibly naive.

He came into the office one day saying that he'd heard on the radio that most people only paid the minimum payment on their credit cards each month. 'Welcome to our world'. No one actually said it, but I could tell that everyone was thinking it. And one of his favourite pastimes was to regularly drive around the perimeter of the district, which was a considerable distance I might add, to check that it was all still intact, and then claim the mileage – to pay off his credit card no doubt.

I handed the phone over to the boss who had witnessed the whole debacle.

"Patient spea…"

He was cut off in his prime (ouch).

Pause.

Funny word 'pause' don't you think? It can be spelt four different ways: pause, pores, pours and paws. Each one has a different meaning, yet they are all pronounced exactly the same. No wonder English is a difficult language to learn. And don't even get me started on Paw Paws!

"I'm the chief enviro…"

Pause.

"You'll just have to take my…"

Pause.

"How would you like me to prove…?"

Pause

"If you will just give me a cha…"

Pause.

"Yes I heard the conver…"

Pause.

"But you're talking to me in exact…"

Pause.

The phone was slammed down.

There was complete and utter silence.

Nobody spoke.

Nobody moved.

I don't think anyone even breathed.

The phone rang again.

"Good afternoon, environmental health, how can I help you," the boss chimed in his inimitable Perky – I had no idea where Pinky was – style.

Pause.

"Oh it's you again," he remarked without the slightest surprise or malice in his voice.

Pause.

"You were warned."

Pause.

"In that case I'll pass you over to the officer that you were talking to originally."

"Hello," I said completely composed, even though I wasn't sure what reaction I was going to get from the other end of the line.

"Hello," was his composed response.

"How can I help you?"

"There's sewage spewing down the lane and it's coming from my neighbour's cesspit." Then before I had chance to respond to the allegation he demanded, "What are you going to do about it?"

"And your name is?" I enquired, still maintaining my composure.

"Why do you need to know my name?"

"Because you're the complainant, that's why," I said firmly.

"Why aren't you asking me for the name of the bastard that's causing the problem? Don't you know how to do your fucking job?"

"Firstly, yes I do know how to do my job."

I wasn't willing to discuss how I did my fucking job with him.

"And secondly you've already been warned about your behaviour."

I couldn't believe how calm I was.

I then heard a deep intake of breath, on the other end of the phone, before he proudly announced, "My name is Tarquin..." double-barrelled, hyphenated with a silent p, something or other.

Of course it was.

I'd switched off after the first barrel to be honest even though I knew exactly what I'd like to have done with a double barrel of another kind. All I heard from that moment on was blah, blah, blah, blah – and then finally, after another, totally unnecessary, melodramatic intake of breath, he finished his title pompously with... "Esquire."

You have got to be kidding me. Who in their right mind calls themselves 'Esquire' in this day and age?

I bet he'd been born with a silver spoon in his mouth as well. It was just a pity it hadn't got lodged in there – in the back of his throat preferably.

I was convinced that he'd had some grandiose vision of himself actually being the local country Squire – with all the benefits!

Well from now on I'm going to call you Richard Head.

But not to his face obviously.

"And your address?"

"It's not my bloody ad..."

"Yes it is," I interrupted him this time, "As well as your neighbours."

I was beginning to get quite good at holding my own.

In the end I managed to get both of their addresses from him, as well as his telephone number. So it was true – 'wonders did never cease'.

"I'll call out tomorrow and see what I can sort out," I finally informed him.

"But I want you here now," he hissed straight back at me.

You can want all you like – matey!

"It's a bit late in the day now, and as you've already said, it's been going on for weeks, so I don't think a few more hours will make that much difference, do you?"

By now I was starting to get a tad stroppy.

Besides, it couldn't have been that much of a problem because we hadn't received any other complaints from people living in the area.

The following morning I set off for the remote village of Marshalhales and when I say remote, I do mean remote. It was right at the back of the back of beyond, very close to the edge of Northford district's northern boundary, and it was going to take me easily over an hour to get there.

It was the perfect day for a long leisurely drive (I had to keep to the speed limit) through the picturesque Shropshire countryside. The early morning autumn sun was hovering majestically, like a mystical golden orb, in the perfect azure sky. There wasn't a single cloud to be seen for as far as the eye could see – and way beyond I suspected – and as I left the hustle and bustle of Northford town behind me, I could hear its muted sounds – humming like a hive of drowsy bees – fading with the ever-increasing distance between us.

I had all the car windows fully wound down on that glorious Indian summer's morning and the radio blaring away throughout the journey. I can assure you, however, that I didn't cause a noise nuisance to anyone because there were considerably more sheep than there were people in that neck of the woods. And talking of woods, which I was, the one I passed on my way to Marshalhales was a magnificent blaze of fiery seasonal colour. The bronze, rust and ochre leaves from the sycamore, beech and oak trees – to mention the only three that I knew the names of, but there were others – floated effortlessly on the gentle breeze as they fell to earth creating a beautifully layered patchwork quilt that I was sorely tempted to stop at, get out of the car, and then go running through kicking, but certainly not screaming.

After just over an hour (I told you so) I turned onto the road that led to Mr Head's house. Actually, to call it a road would be a bit of a misnomer. It was more of dirt track really that had more pot holes in it than actual road surface. *I'm going to be lucky to have any suspension left on my car at this rate.* Then suddenly, and without any warning, I found myself hurtling down a steep incline far too fast for my liking as the car bounced, like a big metal bouncy thing, as I continually swerved to avoid the aforementioned pot holes. Then I felt the rear end of the car starting to lose traction as it swung precariously to the left. I desperately sawed at the wheel and the car then swung violently to the right this time and I was convinced that I was going to go into a complete spin. Finally, at the last minute, the car swung back the other way, then in the opposite direction again, and then I came to a grinding halt – at the bottom of the incline – facing in the right direction again. Phew!

Now all I had to do was negotiate the north face of the Eiger, well the ridiculously steep incline that was now facing me. Thank goodness it hadn't snowed.

Not only did the road surface change dramatically, but the terrain did as well. The magnificent woodland had now turned into wild heathland covered in deep purple heather that was randomly punctuated by the yellow heads of wild potentilla.

It was turning into a botanist's dream trip!

When I did finally arrive at Mr Head's house, exhausted – but in one piece - I parked in the lane directly opposite it. I climbed, gingerly, out of the car and then proceeded to straighten out my arms, legs and spine, like a cat stretching after a long sleep, to get the circulation going again after the long, latterly grossly uncomfortable, and nerve-racking journey. I changed into my wellies, collected the drain keys from the boot of the car and then headed off towards the house.

Parked directly outside, in all its ostentatious glory, was one of those 'penis compensation' cars. You know the ones I mean: a 'Ferrorghini', a 'Lambari' or whatever they're called, in banana yellow, polished to a show room shine and with a personalised number plate: A N0B (only joking). It was actually: N0 NOB (still joking). It was hardly the most practical of vehicles to be driving around the area in given the state of the road surface. There was also a brand new Range Rover, in jet black, also polished to a show room shine and also with a personalised number plate which was, at least, a more practical mode of transport.

His, ever so slightly, annoying, rude and arrogant attitude was starting to make perfect sense now.

The house was spectacularly unspectacular. Certainly in comparison to the cars parked outside it. I'm not quite sure what I'd imagined it would look like to be honest, but it was definitely something a bit grander than the featureless 1960s detached, white washed, brick property I was stood facing.

The large garden was turned over mainly to lawn, which looked as though it had recently been mown because it had those bowling green stripes up and down it, with narrow flower borders, containing very minimal planting, surrounding it. I don't think it was particularly a minimalist look they were aiming for; I think they just couldn't be bothered with it all.

I knew the feeling only too well.

He must have heard the car door slam or my footsteps crunching on the gravel path because before I'd reached the front door he'd opened it with a quite unnecessary, and quite camp to be honest, flourish.

I was greeted by a short (told you so) banker – but not in the financial sense of the word – dressed in the recently popularised 'preppy' style of fashion. He wore a petrol blue and white striped – button down collar – shirt which was open at the neck exposing a red silk cravat, under a navy blue blazer that had a red polka dot handkerchief strategically placed in the breast pocket.

Well I thought they were polka dots; they could quite easily have been spots of blood. Perhaps he'd cut himself shaving that morning – chance would have been a fine thing.

His trousers were white flannel, pleated at the waist, with deep turn ups and completing the ensemble leather loafers – probably Italian – in tan.

His mega-trendy short back and sides coiffured hair – with a quiff – was held perfectly in place by a large dollop of hair cream that gave it that salon finish lustre and emphasised his, as fake as he was, tanned face – which incidentally matched his shoes.

He smelt citrusy; as though he'd rubbed a lemon all over his skin.

I was certain that he probably had other fetishes too.

It was probably one of those expensive designer fragrances; 'Christian Saint Gaultier' or 'STNK'; it certainly wasn't a cheap body spray like 'Bobcat'.

He was immaculate.

I, on the other hand, wasn't.

He also looked a right twit – and if you would like to change the vowel you are most welcome to do so, but I don't know what a twot is!

I, on the other hand, didn't.

Worst of all; Richard Head was...... a Yuppie.

Aaaarrrrrrgggggghhhhh!

Enough said.

"Good morning, how are you?" The words were actually sticking in my throat as I said them.

"Man from the council," was his opening remark, as he fastened the brass buttons on his blazer faithfully following the golden rule; top button – sometimes; middle button – always and; bottom button – never, in a bored monotone voice stripped of any intonation. I wasn't sure if it was a question or a statement – and I didn't really care how he was anyway. (See Chapter Ten). It's just that I was 'dragged up proper' and taught to be polite.

"Yes," I said equally unenthusiastically.

"Who is it darling?" a voice quacked from inside the house.

"Man from the council 'Bunnykins'," was his nauseatingly sycophantic answer.

Who has he got in there, a friend of Peter Rabbit?

"You can't come in," he snapped. "Jemima has friends round for coffee and canasta."

He has – it's Miss Puddle-Duck herself.

I did think of asking for an autograph, but in the end I thought better of it. I would have probably only got a webbed foot print, and that could have been from any old duckie.

'Well I much prefer sherry and charades personally.'

(You thought that I was going to say something else beginning with 'sh' then, didn't you? Come on now, be honest – but I don't know how to play shogi).

All the time he was talking to me he was performing the 'Peacock Strut', an old time sequence dance of yesteryear, strictly <u>not</u> ballroom – even though he had plenty in those trousers. He was constantly playing to an imagined audience and posing for photographs that weren't being taken. I was convinced that he'd have narcissi in the garden in springtime.

It was then that I noticed he had a bandage wrapped round his left hand.

"Have you hurt yourself?" I found myself asking the question before I realised what I was doing.

Why was I even asking? I wasn't in the least bit interested, or concerned even. (See Chapter Ten, again).

Perhaps I was secretly hoping that he had hurt himself.

He looked at me slightly perplexed until I pointed to his hand.

"Oh that. It's nothing, just a little prick."

And those were his words, not mine – but how fitting – and as you are no doubt aware, those four little words have more than one connotation. (I knew his brother – Dick Tation).

I had to ask to be shown where the alleged problem was, because it wasn't obvious to me at all. I hadn't been met by a torrent of sewage cascading down the lane, like some diarrhoeal tsunami, on my drive there as it had been intimated.

Having closed the front door he reluctantly escorted me back to the lane where he pointed out the merest trickle, similar to a man's urine flow that has severe prostate problems – and I'm speaking from experience – of sewage meandering leisurely along the edge of it.

Now don't get me wrong, any sewage is a potential health risk, but this wasn't exactly what I'd been led to believe it was going to be.

"It's coming from that lot," Mr Head stabbed his perfectly manicured fingers, one of which had a white gold – or more likely platinum – diamond studded wedding band on it, viciously towards the house next door. "I'm not a complete fool."

Why which part of you is missing? I was dying to know.

"How do you know?" I was dying to know that too.

"Because I do, that's how." He may as well have finished the sentence with the childish refrain 'na na na wee wee'.

Very scientific – I don't think.

"Well I need to be able to categorically prove whose it is before I can do anything about it," I politely informed him. "So what I'm going to have to do is put some coloured dye into your cesspit and a different coloured dye in your neighbours to determine which one is actually leaking."

"But I've already told you it's theirs. Why won't you listen? It's certainly not mine," was his contemptuous riposte.

So he could recognise his own shit then. What an amazing and unusual talent, and what a great idea for a prime time television quiz show – 'That's My Shit'! They could have had a celebrity edition at Christmas raising money for charities such as 'Bowel Relief' – even though that sounds more like a laxative.

"How do you know?" I asked again, only to get exactly the same answer.

"Because I bloody well do, that's how."

Okay, so it wasn't exactly the same answer.

I just gave him one of my looks. Not exactly sure which one – I do have a selection of them at my disposal (or should I say, 'in my toolbox?!'), but whichever one it was, it worked.

"I'm sorry," he gushed in a blatantly phony sheepish tone. "Does my language offend you?"

*No, **you** offend me – period.*

Smarmy git, but I didn't rise to the bait.

We were going round in circles, and not in a good way.

"That argument certainly wouldn't stand up in a court of law," I told him. "So are you going to allow me to put some dye in your cesspit?"

"But…"

"No buts." I didn't want any ifs either. "Will you allow me or not?"

Eventually, though reluctantly, he agreed.

"Can you show me where your cesspit is then?"

"It's over there," he nodded rather nonchalantly. "But I can't possibly go with you dressed like this."

Ahhh, go on, spoilsport – I dare you. It was just a thought.

"You'll have to go on your own."

Don't you worry; I'd much prefer to go on my own thank you very much. It was just another thought.

He then pulled back his left shirt cuff, complete with an ostentatious gold monogrammed cuff link, exposing his shiny Rolex watch and declared pretentiously – full of his own self-importance – "And besides, I have other more important things to attend to."

Yeah, right.

"That's fine," I said, "I'll do what I have to do here and then I'll go and visit your neighbour. I'll be back (without the IMI Uzi – unfortunately) in a couple of days' time to see what's happened."

He mumbled something inaudible, and doubtless derogatory, then sloped off back to the house treading very carefully so as not to get his shoes filthy.

After trudging cautiously over the undulating heathland for a hundred yards, give or take a metre, I eventually found his cesspit and was grateful to have arrived at it without twisting an ankle or ending up face down in something unpleasant. I was also thankful I'd put my wellies on, although I think I'd have been better off with hiking boots and a Sherpa to be honest. I opened the cesspit lid to find it was as full of as much shit as he was – and that was a lot of shit believe me – and could quite easily have been described as 'The Pit of the Perpetual Pong'. I then proceeded to sprinkle some yellow drain dye liberally into it.

Actually I just chucked a large dollop of the stuff in. Poetic licence is such a wonderful thing.

Then as I stood there staring at that foul-smelling vat of excrement – the cesspit, Mr Head had left, remember – I started to contemplate the meaning of life. Well my life anyway. *Is this all I can look forward to? Is this what my life has become – a pile of poo? How did I get to this point anyway?*

I just felt as though I was going through the motions – in more ways than one.

Chapter Two

Having made a cods − that being the northern colloquialism for a complete mess − of my A-Levels in the summer of 1974; I sat them again a year later and passed them.

What a relief.

In hindsight − and what a marvellous thing that is − I probably shouldn't have gone to the pub the night before my exams the first time round. But hey ho, you live and learn. Well I live anyway.

So, having achieved the requisite grades, I toddled off to Yorkshire Polytechnic to read Architecture.

Now for those of you under a certain age I should probably point out that a polytechnic was a tertiary education institution; offering higher diplomas, undergraduate degrees and post graduate education in both academic and professional vocational qualifications such as; law; engineering and of course architecture, that was governed and administered at a national level. After the passing of the Further and Higher Education Act 1992 polytechnics became independent universities able to offer their own degrees.

Got it? Good.

I'd always been fascinated with building things as a child and had a few of the toy construction sets available at the time including; Betta Bilda; Architex (I think that's how it was spelt − apologies if it isn't); Mechano and of course Lego − in the days when the pieces were either red or white.

I wasn't particularly popular with my sixth form tutor for only applying to 'lowly' polytechnics, but I didn't really care. Being amongst the first pupils nationwide of the new comprehensive education system, to sit A-Level examinations, all eyes were on us to prove that the system actually worked. And that was going to be measured by the number of university entries.

Yes, really.

I always thought that the measure of success of any school was how many pupils went on to do what they wanted to do – whatever that happened to be.

Well I wasn't going to play their silly little game, and apart from that, architecture wasn't taught at university anyway – so I didn't really have a choice in the matter.

At this point you may be thinking that I have issues with authority and don't like being told what to do. Well you'd be right, but only if I can see no logical reason for being asked or even worse made, to do something I don't understand the point of.

As a complete aside – and just out of curiosity – can anyone tell me why you read a subject at higher education and don't just study it?

Now that's got you thinking.

Anyhow, where was I?

Oh yes, Yorkshire Polytechnic.

Well it was a big mistake. A BIG mistake! The reason being; I didn't get on with my tutors – it was as simple as that.

Three of them were practising architects in the area. But as far as I was concerned, all three of them needed a damn sight more practice because the buildings they'd designed were bloody awful – excuse my French.

Believe me, I'm an EHO.

They were just featureless boxes with windows. Oh, and doors of course. Why it took them seven years of study to qualify and then come up with such boring designs I'll never know. A child of four could have produced something better during a creative drawing session at nursery.

None of the tutors allowed students any creative freedom whatsoever with their own designs because the tutors just wanted to produce carbon copies of themselves. So after completing the first year, I left.

Some of the other students in my year felt exactly the same as I did but decided to 'carry on regardless' (it really was a farce), only to be thrown off the course at the end of their second year.

Yes, the tutor's egos were that big.

I'd also been offered places at other polytechnics, but eventually chose Yorkshire because of its reputation as the best School of Architecture in the country at that time. Who knows what would have become of me if I'd gone to one of the others.

But there were no regrets as I believed I was destined for much greater things – and then I became an EHO!

21

Having said all that, architecture was never my original career choice anyway. What I really wanted to be, from a very early age, was a lumberjack.

No seriously, it was a pilot. Unfortunately I had that dream shattered after taking the Ishihara colour test, during a routine medical examination, as a child. It was then that I was told I had red – green colour blindness. And it would appear that this is by far the most common form with as many as 8% of men and 0.5% of women being affected – so now you know. Consequently I wouldn't be able to become a pilot, or a supporter of any team with a red and green strip.

Of course I can see that grass is green and blood is red – unless you're a royal of course. I can see traffic lights okay and besides, everyone knows that the top one is red and the bottom one is green. However, this is how it has affected me in the past. Many years ago I bought a pair of, what I thought were, brown trousers only to be told – when I recently gave them away to charity – that they were in fact olive green.

Never mind eh, worse things happen at sea – apparently.

But what doesn't help people with this visual impairment are colour names such as:

Cat's Paw (does that not depend on what the cat's trodden in?);

Joa's White (it's not even white);

Dimpse (I ask you, how is anyone supposed to know what colour that is?);

Eating Room Red (as opposed to which other room);

Elephant's Breath and Mole's Breath (does breath even have a colour and if so, does it differ from animal to animal?).

All these colours, and more, are featured in the Farrow and Ball paint colour chart.

What can we expect next: Ferret Fart, Snail Sneeze and Cougar Cough?

Finally, while I'm still on the subject of colour, why is it that the 'green room' in a theatre isn't necessarily green?

Meanwhile back at Yorkshire Polytechnic: at the end of the year I returned home and got a job as a computer operator. This was in the days when data was stored on large magnetic tapes in a climate controlled room. I couldn't have imagined desk top computers back then, never mind laptops or tablets. Well that's not strictly true, I could imagine tablets actually. I'd even taken some once called a paracetamol.

Computing was only going to be a temporary job, because at that moment in time I no longer knew what I wanted to do when I grew up. And if truth be told, and keep this to yourself, I still don't! But it filled a gap for a year – even though that's not what most people would consider a gap year to be! Then on the day Elvis Presley died – yes I can remember precisely where I was on August 16th 1977 – I was thumbing through the Manford Evening News when I noticed a particularly attractive advertisement. The one next to it said that Salchester Council were looking to appoint a student Environmental Health Officer and not only were they going to pay them a salary, they were going to pay all of their college tuition fees as well.

Well, I knew what an EHO did – so I thought.

I can do that – so I thought.

It was a no-brainer. I applied and the rest, as they say, is history. And this book is my history. Well, part of it at least.

The night before my interview I had a very disturbed sleep.

No jokes about being disturbed, thank you very much.

I'd like to say that I didn't sleep a wink all night, but that would be exaggerating, and I'm not prone to that – much. I just kept mulling potential questions over in my head whilst trying to predict what they would actually ask me, as well as concocting some suitably convincing answers.

My interview wasn't until the afternoon, but I was awake bright and early – even if I wasn't exactly bright-eyed and bushy-tailed. I didn't want any breakfast that morning because my stomach was churning like a washing machine on a fast spin. I couldn't believe how incredibly nervous I was. Did it mean that I actually wanted to become an EHO?

Having completed my ablutions, in record time – without ripping my face to shreds – I got dressed in my recently purchased, especially for the occasion, outfit. This comprised of: a black suit (single breasted – double would have made me look like a gangster) without the waist coat; white shirt (classic collar, button cuffs – I don't like cuff links); plain black tie (with a half Windsor knot – what else?) and, you could see your face in them, shiny black shoes (lace ups, not slip-ons – I have a high instep). I'd heard that black was the new black. I looked in the mirror and hardly recognised the person staring back at me. I actually looked rather dapper, even if I say so myself. I could quite easily have passed for the man at C&A. And I believe they still exist in other parts of Europe. C&A that is, not men. Well obviously men as well. Oh never mind.

Looking back, they must have thought I'd just come straight from a funeral.

Trust it to have been wet and windy on that of all days. That was the weather, I hasten to add, and not me. Although the way I was feeling, it wasn't far from being both.

Because of the atrocious weather, the journey to my interview was fraught with difficulties. My poor little MINI Clubman – my pride and joy and a 21st birthday present from my mum and dad – was buffeted by the high winds; it was blowing a hooley for most of the journey. I had one hell of a job keeping her in a straight line. I was clinging onto the steering wheel for dear life every time there was, what seemed to me to be a ninety mile an hour gust. The windscreen wipers were working fifty to the dozen and yet I could still only just about see where I was going. And then I had to contend with those perennial idiots who were driving far too fast and without any lights on.

In spite of it all, I arrived in plenty of time.

I'm a bit OCD about punctuality.

The offices of the Environmental Health department of Salchester Council were, and probably still are as I've never been back since I left, nothing particularly exciting architecturally.

Trust me to notice that.

They were just a featureless oblong building with some windows and doors.

I really do believe that Prince Charles and I could have some very meaningful discussions about architecture.

They could so easily have been designed by my Yorkshire Polytechnic tutors, or maybe even one of their protégés.

On my arrival I dashed from the car park to the offices trying to dodge the raindrops that were falling on my head (cue song) and surprisingly, I made it without getting drenched in the process.

On entering the building I informed the undeniably pretty receptionist, with her almond shaped eyes, English rose pale skin and long dark tresses tied back in a ponytail, who I was and why I was there. She ticked me off – even though I hadn't been naughty – her list and then politely asked me, with the cutest of lisps, to take a seat telling me that someone would collect me when it was my turn to be interviewed.

Is it just me, or does anyone else find lisps sexy?

I hadn't realised just how low the seat was when I went to sit down. After lowering myself to the point of no return – where I thought the seat actually was – I faltered, big time. It was, however, too late to do anything about it because momentum had got the better of me and I was plummeting to earth, without the aid of a parachute, at an alarming rate of knots. I just collapsed into a crumpled heap on the seat – like a cartoon character does after plummeting off a ridiculously high cliff. I then glanced round quickly to see if anyone had noticed with a look that said, 'I meant to do that' – like a cat does when it falls off something. My bum was almost touching the floor whilst my knees were positioned at eye level almost obscuring my view, unless I opened my legs.

Can you imagine if I'd had a skirt on?

You can?

You need help.

I must admit though, I was absolutely fascinated sitting there watching the comings and goings of the human traffic through the reception area. Along with Desmond Morris I do like a bit of people watching. Sullen joe public – sounds like a country and western singer – trudged in, grumbled about things that didn't really matter and then trudged back out again. I don't know how the receptionist managed to stay so cheerful, or polite. Later, when I was working in the department, I came to realise just how trivial some, if not most, of the complaints actually were. At times I really just wanted to tell some of them to 'get a life', but valued my job too much.

All too soon my name was called out and, after scrambling out of my seat in a rather ungainly fashion, I was escorted up to the first floor by a rather 'matronly' woman – make of that what you will – with a hair sprayed helmet of a hair do, to the office where my interview was to take place.

As I walked into the room I was greeted by a stagnant blue haze that hung in the air like mist over a water meadow on a cold autumnal morning. Remember we are talking about the days when smoking in places of work was both acceptable and legal.

I've always thought that interviews are really strange affairs. They say (whoever 'they' are) that hiring decisions are made somewhere between the first four and twenty seconds of an interview.

How on earth can that be objective?

Okay, so the panel know something about you from your application form, or curriculum vitae, but what specific criteria are they using to make a decision within four seconds of you walking through the door? That's the FOUR SECONDS that have the potential to dramatically change your life – for better, or for worse (for richer, for poorer). All they can possibly assess in that time is how you look and how you smell.

That meant that I looked how they expected/wanted me to – as if straight from a funeral – and I didn't smell offensive. But in comparison to some of the smells I was to encounter during my career, that wasn't particularly difficult.

I have this, I suppose bizarre, mental image that one of the panel is an expert on smell. That they've been on a course to become a parfumier and can tell the difference between a cheap supermarket body spray and an expensive designer fragrance and therefore score the candidate according to which they are wearing.

I told you it was bizarre.

My interview was nothing to write home about, I was going home after it anyway, so I could tell my parents about it then. It would have been much quicker and considerably cheaper than using the postal service too. In fact it only lasted about forty minutes and I must admit, at the time, I didn't know whether that was a good or bad thing. Had I impressed them sufficiently so they didn't need to hear any more from me? Or had they just had enough and wanted to get rid of me?

And I don't want answers on a postcard, thank you very much.

There were three men on the interviewing panel: Peter Bowden, Stan Pollock and Harold Carter – all of them Deputy Chief Officers.

I only mention their titles because Environmental Health Departments were very hierarchical back in the 1970s – as well as being male dominated – and Salchester's was no exception. Now the majority of EHOs are female and many of them are Chief Officers.

At the helm – nice nautical term that – was the Chief Officer. Under him were the three Deputy Chief Officers in charge of the four main environmental health functions:

Food Hygiene – now referred to as Food Safety – and Health and Safety at Work,

Pollution - now Environmental Protection - and

Housing.

Below the deputies were three principal officers for each of the functions, and the pollution principal was hard of hearing – cross my heart etc. So there was no noise nuisance in Salchester. Then below them were three senior officers.

One of the duties of a senior was to check the EHOs hand written letters before they went off to the typing pool. When I qualified, my senior's command of their mother tongue was not particularly good and so they would regularly alter my letters to make them grammatically incorrect and on occasion change my perfectly good spellings to perfectly bad.

Fortunately the typists had a very good command of English and would change them back before the typed version of the letter was checked, yet again, this time by the principal before it could be sent out. I know, it was bureaucracy gone mad.

Finally there were the ground troops: the EHOs, technical officers and support staff. Yes it does seem as though there were more chiefs than Indians, and there almost were, but I don't think it's quite as bad these days – unless of course you know differently.

Now back to the interview.

"Please take a seat and make yourself comfortable," bellowed a voice from behind a 'large imposing' desk. But from whom it came I couldn't tell.

I suitably obliged – I'm like that.

Once I was seated I nervously gripped the armrests of the chair generating the customary white-knuckle look. You would have thought that I was at the dentist or on one of those scarily fast, throw you all over the place, roller coasters.

It was difficult to make out all the features of the panel because they were sat with the window directly behind them and so I only really got their silhouettes.

Were they the three wise men, the three amigos, or maybe even the three wise monkeys?

Peter Bowden was in charge of Food Safety and Health and Safety, and being the tallest of them came over as a real gentle giant. He also seemed to be the most switched on and in his thick Yorkshire accent asked the most delving and relevant questions. He did not, however, wear a cloth cap, brandish a black pudding or say 'ecky thump' – not even once.

Peter eventually became my first boss when I qualified and in all my years in the profession – and there have been a fair few – I've never had a better, fairer, kinder, more decent boss than him (the list of attributes could go on infinitum). If you ever made an almighty cock-up of things he would always back you up to the hilt and support you through it instead of abandoning you – unlike some of the other bosses I've worked for. He would inevitably give you a ruddy good bollocking behind closed doors

27

and tell you to never make the same mistake again, but he would never hold it against you or ever refer to the incident again. He truly was a Prince amongst men.

I could see that Stan Pollock had a handlebar moustache because it protruded slightly beyond his chubby cheeks. However, it was nowhere near as distinctive or half as impressive as Salvador Dali's used to be. Stan was in charge of the pollution section as well as being the student training officer. I wasn't quite sure why he was the latter really, because he had very little interest in students and I hardly saw him during the three years I was one. Whenever I did see him, however, he would always ask me exactly the same question – forgetting that he'd asked it me on previous encounters. "Where's the ligamentum nuchae?" I would always pretend that I didn't know – even though I did – and then he'd look at me, all superior, as if he'd caught me 'hook, line and sinker', and then proceed to tell me where it was.

What a Pollock!

Harold Carter was the Housing Section's boss. I could see he had a bald head because it was very shiny and reflected the light from the window. He also had a terrible, hacking, smoker's cough that interrupted the proceedings from time to time and was a bit of a distraction to say the least. But other than that I couldn't make out anything else distinctive about him.

The first question I was asked was, "Why do you want to be an EHO?"

Well I didn't see that coming.

This was during the bathing costume round. I knew, however, that it was a trick so I avoided the obvious answer that I wanted to save whales.

Midway through the interview Stan Pollock asked me to stand up.

That's a bit fishy, I thought.

What for? I wondered.

But I was far too nervous to ask, so just did as I was told.

Hard to believe I know.

It transpired that he wanted to see how tall I was. *Is there a minimum height requirement that I didn't know about? Surely he must have noticed how tall I was when I walked into the room?*

So did that disprove the four to twenty seconds theory? Or had Mr Pollock just forgotten how tall I was? Working with him later I discovered that forgetfulness was one of his many peccadilloes – so that probably was the reason.

I eventually went on to score 5.8 across the board for technical merit with one 6.0 for artistic impression (old scoring system) which, I must admit, I was pretty chuffed with.

Anyway, the upshot of all of this is: I got the job. But I'm sure you've already worked that out for yourself anyway.

Student training in the late 1970s was vastly different than it is today. In my day – and I always vowed that I'd never use phrases like that, but as I get older, although not necessarily wiser, I find myself saying and doing things that my parents used to – the course was a 'thick sandwich'. That's nothing to do though with a northern butty made from dockers wedges i.e. very thick hand cut bread. In other words, students would spend a block of say two months at college and then perhaps three months back with a local authority to gain practical experience. That's if they were lucky enough to have a sponsor, because many didn't. This pattern was repeated throughout the three year course, although the length of the blocks would vary. All college holidays were spent on placement.

These days, students spend their first two years at university, then their third year out with a local authority to complete their practical training logbook – a bit like a National Vocational Qualification (NVQ). That's if they can actually secure a placement because gaining practical experience is still a problem because so many local authorities don't offer placements. All I can say is; 'Shame on you – you know who you are.' That's then followed by the fourth year back at university to complete their degree.

Some students opt for end on training, so they complete their degree in three years and do their practical training in their fourth.

Needless to say I think my training experience was much better. That doesn't mean to say that there weren't problems with it, there were, many and varied. But at least I could get the practical experience of the theory I had recently been taught – notice I haven't said learnt – at college, sooner rather than later.

It's similar to when you have driving lessons. You're taught how to drive a car to pass the test, but actually only really learn how to drive when you have passed and are out on your own. The same applies to environmental health, as I'm sure it does to many other jobs.

This regular mixture of taught theory and practical on the job training helped to make sense of it all that much quicker, although that wasn't always the case. As we all know, only too well, the theory and the practice can be vastly different, as I found out when I eventually qualified. But I'm getting way ahead of myself.

Salchester College had a reputation for being the best in the land – that's 'Neverland' I eventually found out – at that time to study Environmental Health.

Have you noticed that there's a pattern starting to emerge here?

The reason for that was the pass rate was high in comparison to the other colleges that offered the course. Unfortunately, Salchester College's reputation was completely unfounded. The main reason was that a lot of the lecturing staff were not particularly good. In fact some of them were absolutely useless to be honest and as a result we had to do a lot, and in a few cases most, of the work for ourselves for fear of failing. Not a bad thing I know, but you at least expect the lecturers to be imparting up to date information. Some of them were using their notes from when they were at college back in 1874. A slight exaggeration perhaps – even though I'm not prone to them – but it certainly felt like that. These notes often included references to legislation that had long since been repealed and standards that were no longer used.

One lecturer, who taught 'Food', all Tuesday afternoon, in the final year, spent hour upon hour describing a veritable harvest of fruit and vegetables and included the classic line, "Peas are green and spherical."

Well I certainly hope there's a question on peas in my final exams, because I think I'll be able to answer it. It's information like this that every EHO needs to know.

So as a result, excuse the pun, of our own efforts, the pass rate remained high.

Chapter Three

The centre of Salchester was not only an architectural, but also a town and country planning, disaster. Most of the lovely old terrace houses had been demolished and replaced with ugly new, prefabricated concrete and glass, high rise blocks of flats – or 'streets in the sky' as the planners liked to refer to them. 'High-rise Ghettos', were what they'd actually created and in the process destroyed the community spirit that had existed, for so many years, in the so-called 'slums' of the area.

I realised that the houses had become unfit for human habitation; well I did when I started inspecting similar properties and seeing the condition they were in. I understood that clearance areas needed to be declared, but why not replace like with like? And no, I don't mean unfit with unfit, but new fit houses for the old unfit ones?

It was to one of the aforementioned monstrosities that I went with Mick Lacey on my first (of what was to become many), food poisoning investigations.

Mick was one of the few EHOs at Salchester that I actually looked up to. Not because he was taller than me, because he wasn't – in fact he was at least three inches shorter – but because of his positive attitude and unequivocal professionalism. He'd only been qualified for a year when I started as a student and so was still in the honeymoon period of his career. He was unbelievably, as well as undeniably enthusiastic about the job and eager to try new and innovative approaches to solve the old perennial problems. This was in stark contrast to some of the more seasoned officers who were not only long in the tooth – literally in some cases – but were also way past their 'best before' date. Over the years since qualification they'd become increasingly apathetic and set in their generally unsuccessful ways.

He was extremely conscientious too, and very methodical in the way he thought and worked. If a tidy desk was the sign of a tidy mind then Mick's mind was most definitely ship-shape and Bristol fashion.

Physically there wasn't much to say about Mick as he had no unique or distinguishing features. Facially he had two eyes, which were both the same colour – unlike David Bowie's. He had a nose and a mouth as well, but there weren't any warts or moles adorning his face. He also had hair, but it wasn't dyed bright pink nor did it hang in Rastafarian dreadlocks, it was just fair and short. He did, however, wear a stud earring in one of his ears – but I can't remember which – which was extremely radical for the time and something the 'upper echelons' were not very happy with, but could do diddly-squat about.

He dressed very professionally too. Not too formal – in a way that could intimidate people – but in more of a 'smart casual' style I suppose you would describe it. He always wore plain, no stripes or herring bone, trousers with a single perfect crease pressed firmly down the centre – no tram lines for him – and a contrasting sports jacket. His shirts were consistently expertly ironed and there was never a frayed collar or cuff in sight.

He always gave the distinct impression of being perfectly calm, even when under pressure, and had the patience of Job when it came to explaining things – calmly and with perfect diction – to the public, business owners and me. I never heard him raise his voice in anger once – although he could be very assertive when he had to be – and I was never aware of him 'throwing a sickie' and then blaming it on the stress that the work inevitably generated.

He was the perfect role model.

The way we found out about this, and in fact all the other suspected cases of food poisoning, was via a doctor's surgery. Doctors were under a statutory obligation to notify the Proper Officer of the local authority if they suspected a patient of having food poisoning. They even got paid – quite handsomely I might add – for every notification they made.

The role of the EHO, in this instance, is to investigate possible sources of the illness and to instigate action necessary to prevent further cases occurring. And with that very much in mind, Mick and I set off to visit the department's most recent notification, a Mr Grimshaw, and using our trusty A – Z map book found his address quite easily.

It will come to you as no surprise, I'm sure, that when we arrived at the flats the lift was out of order. Not only that, but the entrance hall – which was decorated with some beautifully artistic graffiti – I kid you not – had that distinctive aroma of 'unflushed urinal' about it. And it wasn't from a plug-in air freshener either.

Mr Grimshaw, naturally, lived on the ninth floor. But it could have been worse I suppose, it could have been the top floor. In which case I think we'd have both turned straight back round and hot footed it to the office.

I was relatively fit in those days – relative to what I'm not willing to tell you – but even I was wheezing by the time we'd reached the ninth floor. Poor Mick was positively gasping for breath and there weren't any oxygen masks to fall from an overhead panel either, so at the risk of sounding like a couple of perverts making an obscene phone call, we caught our breath before we rapped – but not in the speaking or chanting rhyming lyrics way – on Mr Grimshaw's door.

After what seemed like an eternity, the door was finally opened and we were confronted by a rather dishevelled, skinny and gaunt-looking man who did not look the picture of health. He reminded me of 'Riff Raff' from 'The Rocky Horror Show'; except Mr Grimshaw had thick, matted hair, that he kept scratching, and his face was buried deep in a bushy beard that was decorated with a conglomeration of food debris and other unidentifiable items. Bogeys it wouldn't have surprised me.

Okay, so he was nothing at all like Riff Raff.

He also had a whopping big carbuncle (I didn't know where Simon was) on the side of his nose that looked ready to burst and discharge its putrid contents indiscriminately. I just hoped that it didn't erupt while we were there. I'd read what had happened in Pompeii and I didn't want to get caught up in his lava flow.

Mr Grimshaw wore a pair of grubby old denim jeans that looked at least two sizes too big for him. *Surely he hasn't lost that much weight with the illness.*

He also wore a shabby pullover with umpteen holes in it – holes that were obviously not part of the original design. It also had a variety of indefinable and probably 'stubborn' stains embellishing it. I doubted even the most powerful of biological soap powders, on a hot wash, could have tackled them. There was also a recent smear of snot down the left arm that glistened like a slug trail on a garden path in the early morning sunrise of a July morning.

Just so you know; food poisoning is any illness caused by the consumption of contaminated food, or drink, and is more common than most people realise.

The common causes are:

(a)   microorganisms i.e. bacteria (or their toxins), viruses and moulds.

(b)   poisonous plants

(c)   allergies i.e. nuts

The symptoms include:

(a)   abdominal pain

(b)   diarrhoea

(c)   nausea

(d)   fever

(e)   vomiting

(f)   collapse

Sound familiar? And no, I don't mean how you feel after those ten pints on a Friday night.

Mick explained who we were and why we were there. No identification was asked for, and even though we volunteered it, it was only given a cursory glance.

"Can we come in please?" Mick enquired.

"Ugh," was his Neanderthal response.

Mr Grimshaw's flat consisted of a long narrow corridor from which the bedrooms, bathroom and kitchen came off and at the end of it was the living room. As we entered the flat Mr Grimshaw and Mick set off down the corridor leaving me to close the front door. When I turned round I couldn't see a thing. It was pitch black as all the doors were shut, and all I could hear was a strange slurping sound. However, not wanting to appear scared (I was actually bloody petrified – I'd seen the film 'Alien') I started to walk calmly towards the strange slurping sound.

'Don't do it,' I hear you cry – I have incredibly acute hearing. 'You're heading towards the monster.' But I had very little choice. It was either that or stand there like someone who was afraid and just standing there; and I'd already done that.

Fortunately I didn't have to wait long before the living room door was flung open and subdued sunlight streamed down the corridor – highlighting a multitude of dust motes that danced randomly on the turbulence of the recent air movement – and I could see where I was going at last.

The living room looked in desperate need of a makeover. If only there was a television programme that did that. Unfortunately 'Changing

Rooms' didn't appear on our screens until 1996 and I don't think MDF had even been invented then!

The walls were covered in what I imagined was once a bright orange, but now severely faded, blousy geometric patterned paper that was peeling off around the edges. It must have been an absolute nightmare to match up and with a lot of wastage on each drop too.

The windows were devoid of curtains. I guess he thought he didn't need them being so high – in more ways than one as it turned out. They also had colonies of black mould in the corners that looked like mounds of caviar – Ossetra not Beluga, obviously – which were the result of a chronic condensation problem.

I don't think he'd discovered the benefits of ventilation.

The ceiling had an impressive collection of cobwebs dangling from it that Quentin Crisp would have been proud of along with a selection of fly papers that had a number of the pesky insects stuck to them in various stages of decomposition.

The room was also sparsely furnished. There was an old Ferguson television set – which you had to get up to, to change the channels and alter the volume – quietly talking away to itself in the corner. There was also a decrepit leatherette armchair, of indeterminate colour, with a collection of cigarette burns gracing the arms. This was strategically placed right in front of the TV – so that he didn't have to get up to change the channels or alter the volume I presumed. But the room was mainly dominated by a 1960s white tubular steel and – badly scratched – red laminate kitchen table with four white tubular steel and red plastic padded chairs. The table was covered with a film of grease that; not only could you have engraved your name in, but John Travolta and Olivia Newton John would have been embarrassed by. It also had a plate on it with the remnants of what I presumed was a previous meal along with a rusted can of what, according to the label, was once fruit cocktail, and a desert spoon sticking out of it resembling an oversized cocktail stick. There was also a cereal bowl, full to overflowing, with a selection of 'cigarette' dimps in it.

How old the leftover food was would have been impossible to determine without the aid of carbon dating, but I reckon Alexander Fleming would have been able to find the mould on it that eventually led to his discovery of Penicillin.

There was also a large chipped mug on the table which had a lining of coke (black not white) – from the numerous drinks of coffee that had been consumed from it without being washed out – which must have reduced the capacity of the said mug substantially.

I dreaded to think what his kitchen looked like and I suggest you get that vision out of your head, pronto.

I don't think he was living in abject squalor, but he was certainly living next door. And if cleanliness was next to godliness, then he must have been an atheist.

After introductions and pleasantries were exchanged, Mr Grimshaw grunted, "Wanna seat?" through a mouth full of broken, discoloured and decaying teeth which along with the 'tobacco' gave his breath a distinctive sour aroma. Fortunately we noticed just in time – with our backsides hovering precariously over the chairs – that there were in fact no seat pads, just gaping holes where they should have been. So, unsurprisingly, we declined the offer.

I could **still** hear the strange slurping sound.

Even though cleanliness wasn't at the top of Mr Grimshaw's list of priorities – it wasn't even half way down – I have to admit that courtesy certainly was. His offer of a seat was quickly followed by, "Wanna drink?"

I did want to scream; 'YOU MUST BE BLOODY JOKING,' but refrained from doing so as I didn't want to embarrass Mick, offend Mr Grimshaw (as if) or show myself up. I could quite easily do the latter without the need to scream and/or swear.

We both graciously declined the drink as well.

I could **still** hear the strange slurping sound.

Mick explained to Mr Grimshaw that we needed to ask him a few questions. "We need to ask you a few questions," Mick explained. "We need to know what you've been eating recently, where that food was eaten – at home or away from home – and where you bought your food for home consumption from. Is that okay?"

He wobbled his head like an Indian does when asked a question in that indeterminate yes/no fashion.

Because it had taken so long for the doctor to send Mr Grimshaw's notification through to us we were on a hiding to nothing because he couldn't remember what he'd eaten the previous week when he'd been to see his doctor. In fact he couldn't even remember what he'd eaten the day before our visit, if he'd actually eaten anything at all.

The main reason for asking these, rather intrusive, questions is to establish if individual cases of food poisoning are actually part of a larger outbreak, i.e. unrelated people ate at the same restaurant on the same night and had the same food and everyone was made ill. It's much easier to establish an outbreak at, for example, a wedding reception where sixty out of the eighty guests, who ate the same food, are made ill.

Having said all that, statistically more cases of food poisoning actually occur in the home than anywhere else. And that makes perfect sense when you come to think about it because more meals are prepared and eaten in the home than anywhere else. That's no consolation when you have to tell someone that they've probably poisoned themselves though. As you can imagine, that particular bombshell usually goes down like a lead balloon – which would be a great name for a band.

He was also asked about his occupation, because if he was working in the food industry he could quite easily pass on the bacteria, that may have caused the illness, onto other people. Unfortunately for him, but fortunately for us, he wasn't employed and therefore would not have to be excluded from work if he was a food handler and his sample (see below) came back positive.

It was unlikely that he'd been travelling abroad and picked up the illness while away, but we had to ask anyway.

"Have you been on holiday lately?" enquired Mick.

"What d'you fink?"

We took that as a no.

It was difficult to tell if he had any pets, certainly there wasn't a fish tank or a bird cage in evidence. However, there was a rather pungent odour, engulfing the whole of the flat, which on entry grabbed you viciously by the throat and wouldn't let go. And it wasn't just the perfume of the great unwashed – although that was certainly a contributing factor. It transpired that he didn't have any pets, so infection from that source was another dead end.

While Mick was asking the questions and I was making notes, I continued to suss out the joint – and there were plenty of them in the cereal bowl believe me. During this time I noticed that Mr Grimshaw was continually and rather nervously pacing up and down the room as he was talking to us and the strange slurping sound was following him around. When I eventually looked down at the floor I couldn't believe my eyes. To my shock, and horror, it became perfectly clear what the strange slurping sound was. Everything that had ever been spilt on the floor must have still been there. He was walking around in his bare feet and the gunk was oozing between his toes and then making the strange slurping sound when he lifted his feet up after being stuck to the floor.

Nice!

It beggared belief that anyone could live like that at the end of the 20th century. Had he not heard of the new fandangled craze: cleaning with soap and hot water? A rhetorical question obviously.

At this point even I, as a wet behind the ears student, had worked out that he'd probably made himself ill in some way and was extremely grateful that I'd refused that drink.

This was something that, throughout my career, I always found quite amusing. It was always guaranteed that when I was in a 'dirty' premise, I would be offered refreshments but never when in a 'clean' one. Likewise, when I was in a 'dirty' premise I would be bursting to go to the toilet, never when in a 'clean' one. The law of sod I believe.

I know it was probably psychosomatic, but I now started to itch – like a flea-bitten mongrel. I was even starting to imagine blotches appearing on the back of my hands – no not hair on the palms of them. I was convinced something was making a meal out of me and so started to scratch myself discreetly.

Surely he hasn't got lice as well? Please no.

At last, it was my turn in the proceedings. I had to explain that we needed to get a sample from him to determine if it was indeed food poisoning that was causing his illness and if it was, which type he had. I then produced a metal pot, but not in the way a magician does as it was in my pocket all the time.

Oh, I forgot, that's exactly how a magician does it!

The pot was similar to those canisters that 35mm film for an SLR (single lens reflex) camera came in – in the days before digital photography. You remember, when you used to post your film to be developed and then wait patiently for your photographs to come back whilst singing; 'someday my prints will come'.

By the way, the pots are now plastic. Thought you'd like to know that.

I then explained that we needed a sample of his motions. I stressed that we didn't require it there and then; we would leave the pot with him so he could produce something at his leisure.

A blank expression just sat motionless (which is what we were going to be at that rate) on his face.

So he doesn't know what I mean by motions, okay, let's try, "Number twos?"

Still nothing.

Finally I told him we needed some poo in the pot and preferably his own.

Halleluiah, at last we were cooking on gas – but not the poo I hasten to add as that would have been both unhygienic and extremely offensive.

Would we have been able to tell the difference?

At least I didn't have to resort to the word shit. So I suppose I should have been thankful for small mercies – but I wasn't.

I pointed out that we would prefer a fresh sample and asked him if he was regular and if he thought he would be able to manage one the following morning. (This was all good practice for when I would be obsessively discussing my own bowel movements – and anybody else's who cared to join in – when I eventually became a geriatric). I even described how to get the poo in the pot, because believe me it's not that obvious as the pot only has a narrow neck.

"First, lay a few sheets of toilet paper across the water in the toilet bowl," I explained slowly and deliberately, as though I was talking to someone who didn't understand English. If he still didn't understand me, the next step would have been to shout the instructions. I assumed he used toilet paper, but then remembered that I shouldn't assume anything.

"Next, poo onto the toilet paper and then quickly, before it sinks, use the wooden spatula inside the pot – bit like the one you get with an ice cream tub at the cinema – and spoon some of it into the pot." You'll never look at chocolate nut sundae in the same way again!

There's no need to show off and fill the pot, a small amount will suffice.

I didn't actually say the last sentence. But perhaps I should have.

"Have you got that?" I asked apprehensively.

"Yes," he declared, probably a bit too eagerly.

Job done, or so I thought.

Mick looked at me with a knowing smile.

What did I know!

We wiped our feet as we left the flat.

The following morning Mick asked me to go back to Mr Grimshaw's to collect the pot and then bring it back to the office so that it could be sent off to the public health lab, at the local hospital, to be analysed.

Can you imagine doing that all day? Can you imagine telling your careers advisor that that's what you wanted to do all day? Can you imagine having that listed as your occupation in your passport (pre 1982)? Can you imagine saying what you did for a living when asked at a party?

The lift was still not working, but worse than that some moron – not from the Latter Day Saint movement – had spray painted their name all over the artistic graffiti I told you about earlier, which I thought was a crying shame. Obviously 'Dylan' was seriously envious of their rival graffiti artist's talent and needed to destroy it.

The previous day's odour was, however, still in evidence – perhaps even a tad stronger.

Again it took an age for Mr Grimshaw to open the door and when he eventually did, he looked at me as though I'd just arrived from another planet. I wouldn't have minded, but I wasn't green (as if I'd have been able to tell), my ears weren't pointy and I didn't have a television aerial sticking out of my head either.

He looked exactly the same as he did the day before; dishevelled, skinny and gaunt, and he still had exactly the same clothes on as the day before too – including underwear it wouldn't have surprised me. The only difference was that this time the carbuncle on the side of his nose had burst and there was now a yucky yellow pussy smudge down his moustache and beard that reminded me of 'The Mallen Streak'.

It was obvious by his expression that he didn't recognise me, so calmly and carefully I explained who I was and why I was there once more and eventually the light came on – but only at the speed of an energy efficient bulb.

He invited me in again, but for obvious reasons I made some lame excuse why I couldn't and asked him for his poo pot.

Unfortunately not only had he forgotten who I was, but he'd also forgotten about the pot as well. I didn't bother to enquire if he'd remembered to poo that morning or not. It was completely irrelevant now anyway.

"Not a problem," I assured him through gritted teeth. "I'll return tomorrow if that's convenient for you."

He agreed that it was and I went on my merry way.

Well actually I just left, but 'merry way' sounds so much more poetic.

This fiasco went on for another two interminable days.

I was convinced that I was trapped in my very own Groundhog Day.

No wonder Mick gave me that knowing smile; he'd been there, done that and got the t-shirt. He probably had a drawer full of them.

Finally on the fourth day we hit the jackpot, and I'm only using that as a metaphor because what he presented me with was certainly not a prize I would have wanted to win, believe me.

As he came out of the bathroom he clung on (is that a dead Klingon?) to the pot for dear life with his bony nicotine-stained fingers before he finally unravelled them and proudly presented it to me. The sight I was greeted with was stomach-churningly disgusting. How I didn't show him what I'd had for my breakfast that morning, I will never know. There must have been more poo on the outside of the pot than there could have ever been inside it.

Aren't you glad this isn't a scratch and sniff book?

It looked as though he'd managed to get some under his fingernails as well.

I wish he'd realised that I wasn't just doing this for shits and giggles, although to be fair, there was plenty of the former even if the latter were conspicuous by their absence.

I can't remember the exact words I used at the time, but I did, somehow, refrain from using any expletives. Well, at least I think I did. As you can imagine, I categorically refused to take the pot in that condition. I can only think that he'd held the pot under his bottom (tempted to use another word there) as he was having a poo to get it into that state. I have no idea if there was any poo in the pot, and I wasn't willing to find out either.

So we were back to square one. Do not pass go, do not collect £200.

I told him I'd return the following week with a new pot and we'd start from scratch – which was probably how he'd got it under his fingernails in the first place.

I did eventually get a sample from him. Well he said there was something inside the pot and the outside of the pot was clean, so I took his word for it. I wasn't going to look inside that was for sure.

He must have been telling the truth because when we got the results back from the lab, they were negative for food poisoning organisms and they can't say that if there was nothing in there to analyse. This was hardly surprising given the length of time between him going to the doctors with the symptoms and us getting a useable sample from him.

When I called round to tell him we didn't know what had caused his illness, he didn't really seem particularly bothered.

Who said it was a crap job?

Chapter Four

It would appear that the one thing 'Elf Inspectors' – and by elf I don't mean the supernatural beings from Germanic mythology and folklore – are famous, or should that be infamous, for is inspecting food premises. And that's every conceivable type of food premise: from cafes to hotels, supermarkets to burger vans and everything else in between.

It would also appear that the world and its mother are desperate to hear any juicy stories that are associated with the inspections too, and in my – dare I say considerable – experience, the juicier the better. The public just can't get enough of all the disgusting and repugnant things discovered during inspections, even though they don't really want to know about the premises they frequent.

I think it's what's known as a 'morbid fascination'.

My first ever food hygiene inspection, as a student, was quite uneventful and somewhat forgettable to be honest because the premises were actually 'up to scratch'. In fact the only reason I do remember it is because it was so forgettable, if that makes sense. But as Pat – the qualified officer who accompanied me – quite rightly pointed out, "It's a very good place to start," whilst she gently strummed her acoustic guitar and swayed rhythmically from side to side.

Pat Day was a cocky newly qualified EHO when I first met her and I must admit I actually found her a teeny-weeny bit scary.

As a person she did have a heart of gold, but as an EHO she was a completely different kettle of fish.

To be honest, she didn't do herself any favours. She had a relentless penchant for power dressing, and although Dynasty wasn't on our screens for another two years, dark (tight fitting) shiny satiny tailored suits – that probably were satin – with enormous shoulder pads and teeteringly high, high heels were the order of the day. Her hair, more often than not, was scrapped back severely off her face into a ballerina-type bun that perched on top of her, flat faced, head. She looked like a wannabe dominatrix.

Not only did this elicit ridicule in the office – yes bullying was alive and well in the 1970s – it must have been really intimidating for the unsuspecting public/proprietors when she turned up knocking at their door.

Her choice of language wasn't always particularly appropriate either, and I don't mean that she regularly swore like a trooper. If that had been the case, then I think more people in Salchester would have warmed to her and maybe even understood her better.

I can remember one occasion, quite distinctly, when we were inspecting a butcher's shop in the run down shopping precinct in the centre of town. It has to be said – quite a rough area. The quality of the meat on sale did not, in any way, rival that in Waitrose, and needless to say the shop wasn't very clean either. It was when she balled at them, full blast, in the 'packed to the rafters' shop, "There are anaerobic spore bearing rods all over that chopping board", I knew she'd lost it, and them. Even I was straggling lamely behind. I'm sure they saw it as some kind of compliment and would use it in their forthcoming advertising campaign and as an excuse to increase their prices. 'Our meat contains anaerobic spore bearing rods – buy one, get one free'. The one thing I did learn pretty quickly was that you had to use appropriate language to the receiver. There is no point in telling them something if they don't understand what you're going on about. And that is what is known as stating the bleeding obvious.

In complete contrast to my first inspection, one of the most memorable from that period of my career – and one I will never forget – was to a factory canteen on the outskirts of Salchester. It wasn't memorable because it was a factory canteen per se, but because it was the first inspection I'd taken the lead on – well, for a short time at least – and because of what actually happened during it.

After about twenty minutes of me commencing the inspection Pat had to take over from me. Not because I was incapable – or incompetent (or incontinent) – but because of what I discovered.

I was prepared – even though I'd never been a boy scout – for the inspection to take much longer than the average one because of the size of the kitchen, but I wasn't prepared for what lay in store for me in the actual store.

Inspections were, and still are I presume, always carried out unannounced – for obvious reasons. We didn't want proprietors hiding any incriminating evidence or cleaning the place up just because they were about to be inspected. We wanted to see how they functioned 'normally' on an average daily basis instead of them giving the false impression that they always complied with the law.

Some of the more cynical of you reading this might say that we were just trying to catch them out. Well my response to that, in my humble opinion, spurious accusation would be; If they were doing exactly what they were supposed to be doing, then we wouldn't be able to 'catch them out'. Would you not agree?

I knew you would.

When we arrived at reception, we politely introduced ourselves to the somewhat scatty looking and rather flustered receptionist whose name, according to her name badge, was Beverley. We presented our ID; stated the purpose of our visit; and then asked if we could see the catering manager.

"Have you got an appointment?" she stuttered – now in a complete flap.

"No we haven't," replied Pat – not in any kind of flap. She was as cool as a cucumber that had just come out of the fridge.

"I'll call him and see if he's available."

She then started to clumsily thumb through what must have been the internal telephone directory. She obviously didn't know his number off the top of her head. A head that had a very becoming, hand knitted, purple beret placed at a jaunty angle on it – very trendy, somewhere! I guessed that she was new at the job; it might have even been her first day.

By the time she'd found the number she was in utter panic mode. I don't know if that was because of who we were, or because she just had difficulty meeting the public face to face. If it was the latter, then she was most definitely in the wrong job.

Having dialled the number, she proceeded to talk in very hushed tones – as if what she was saying was some kind of secret – even though there was no one else in reception, besides us, to overhear her. She then nervously informed us that someone would collect us shortly.

We waited, and waited, and waited, until someone finally came to collect us.

Not the greatest first impression.

You're always, justifiably, suspicious that that type of ploy is a delaying tactic because someone somewhere is busy hiding all the stuff that shouldn't be in the kitchen whilst others are frantically trying to clean the days/weeks/months, or in some cases even longer, worth of crud that has accumulated and been ignored up until that moment.

When the catering manager finally did greet us he had an obvious look of fear in his, too close together – he was almost a Cyclops – monobrowed, eyes.

Not a good sign – there was obviously trouble at t' mill.

Now some people would say that having eyes that are close together means that the person is untrustworthy. Perhaps we were about to put that theory, or more likely 'old wives' tale', to the test.

Mr Lockwood's – the catering manager – too close together eyes made him look as though he was boss-eyed, but I don't think he was, and his extremely high forehead – allegedly a sign of intelligence and prudence (seeing ahead), I reckoned that he knew he was 'up shit creek without a paddle' – made him look as though he had a receding hairline. But I don't think he did.

After nervously greeting us he led us slowly from the reception area to the kitchen, which was a distance of approximately 500 metres. It took just under 12 minutes to complete the short walk, which equated to a speed of 2.5 kilometres per hour and is just half the average walking speed for a human. (We were actually overtaken by a tortoise and two snails). All the time he was walking he was uttering utter rubbish – not a unique skill, because I can do that as well – or jabbering utter rubbish would be a more accurate description.

When we all eventually congregated in Mr Lockwood's office, which was just outside the kitchen, I carefully explained what I would be doing during the course of the inspection. This didn't seem to alleviate his 'fear' in the slightest. If anything it seemed to heighten it. I was now starting to get more than a little bit worried about what I'd let myself in for.

Before we entered the kitchen we garbed up in our clean white protective 'lab coat' and white catering trilby hat – very debonair, although some would say 'very last season'. We then got our thermometers out – which, I hasten to add, is not a euphemism – and made our way into the kitchen.

As we entered all of the staff slowly turned their head's away from what they were doing and stared at us menacingly like zombies in a Stephen King novel.

I started the inspection by looking inside the fridges to see if the food was being stored correctly i.e. raw food at the bottom and cooked food at the top, as well as checking that they were at the correct temperature.

Then all of a sudden I caught a fleeting glimpse, out of the corner of my eye, of something hairy scurrying along one of the work surfaces. When I looked around everyone was getting on with their work and no one was flinching. Perhaps it was just my imagination (running away with me).

When I moved to the pastry section there was a blue rinsed middle aged woman with a tight curly perm, electric blue eye shadow and hot pink lipstick, diligently rolling out pastry with a wooden rolling pin. Nothing at all unusual about that you might think. Well, technically what she was doing was illegal, because untreated wood shouldn't be used in a commercial kitchen. However, that was a minor offence in comparison to what I was about to observe.

As she was rolling out the pastry it wasn't going completely flat, as you would expect it to, because part of the rolling pin was missing. When I asked to have a closer look at the pin I could see that there were in fact teeth marks in it. Not just a few either, there were a lot. It was obvious that they weren't made by a human either, like the ones I'd been shown in the photo during my final exams – I was not looking at a case of pica this time – no; these were most definitely the teeth marks of a rodent, a rat in fact. And more than likely it was *Rattus novegicus*, the common rat. So the reason the rolling pin didn't roll out the pastry flat was because the rat, or more likely rats, had been gnawing at it – and for some considerable time by the look of it. I couldn't believe that they thought it was acceptable to be still using it in food production.

This evidence alone didn't prove that there was an active infestation in the premises, only that there had been one sometime in the past. But I was now like a 'rat up a drainpipe' – to coin a phrase – and rushed round the kitchen 'like a headless chicken' – to coin another – to see if I could find any evidence that it was a live infestation.

I got down on my hands and knees to look under cupboards. I moved anything and everything that was moveable so I could take a good look behind and eventually found the evidence I was looking for, rat droppings – plenty of fresh rat droppings.

So how did I know they were rat droppings? Because they were dark brown – naturally – and tapered at one end as well as being spindle shaped and resembled a large grain of rice.

How did I know they were fresh? Because they were moist and glistened, as well as being squidgy to the touch. Old droppings would have been dull, dry and flaky to the touch.

So it was indeed a live infestation. But where were the rats?

Come out come out wherever you are.

Again, as I rummaged around, I was convinced that I saw something move across the work surface to my right. When I looked round, slowly this time so as not to frighten the mystery visitor, I could not believe my eyes. There was a whopping big rat walking nonchalantly along the work surface, as if it owned the place, and no one was batting an eyelid, least of all the rat.

It was blatantly obvious that the infestation was a long-standing one because of: (a) the number of droppings I found – not the exact number, I didn't count every single one, but there was a hell of a lot for just one rat to have produced – and (b) the attitude of the rat – soon to be rats.

Now rats are notoriously shy creatures and usually most active between the hours of dusk and dawn, which usually helps to keep them undiscovered for a long time in an infested premise. They are generally much more afraid of humans than humans are of them. Honestly.

This rat, however, wasn't in the slightest bit bothered by the humans, and peculiarly the humans didn't seem to be bothered by the rat either.

There was also a spotty young lad working in the pastry section who was busy making the pudding for the lunchtime service.

"Is that Spotted Dick?" I asked, 'day glow' woman.

"No," she said looking quite puzzled, "That's Chris, and he's got acne."

Joking aside though, I was seriously concerned about what was actually going into the pudding. Were they in fact dried currants or, the freely available, and in vast numbers, fresh ones – if you get my meaning?

By this time Pat had joined in the hunt for more evidence and when we eventually went into the store, we discovered the rat's nest right at the back underneath the bottom shelf. The store was no longer used for storing food, it just housed; a load of obsolete equipment; a couple of filthy bikes; and some rusted oven wear – and of course, the rats. Just because they didn't use it for food anymore didn't mean to say that it shouldn't be checked, and cleaned, on a regular basis.

It was plain to see that the premises would have to be closed down. Not that I would be able to do that, I wasn't authorised to do it because I wasn't qualified yet, so Pat would have that dubious pleasure.

Unfortunately, in those days, it wasn't as easy to do as you would think. The owner of the business had to be given three days' notice to appear in court where the application would be made, and evidence given, to close the premises.

Now just take a moment to think about that. If an officer truly believed that there was 'an imminent risk of injury to health in food premises' to quote the official legalese, i.e. the general public were at risk of being made ill by eating the food prepared and served at a particular premise, then why did it have to take so long to close them?

Think of it this way. Take one of the large burger restaurants – the choice is yours – and consider the number of people who go through their doors in just one day, never mind three. If there was an imminent risk of injury to health there – and I'm not suggesting that there ever has been, or indeed ever will be – then how many people could potentially be made ill before it was closed? Basically EHOs were expected to do their job with one hand tied behind their back – and their dominant one at that – as well as wearing an eye patch. But at least the choice of eye was theirs.

Eventually the government saw sense and recognised the absolute stupidity of the situation and when the, then new, Food Safety Act came into force in 1990 all that changed. EHOs were given, what is known as, delegated powers, that enabled them to serve a notice, the effect of which would be to immediately close the premises, or part thereof, on the spot, there and then, and straight away. The procedures to serve these 'hygiene emergency prohibition notices', as they are called – which then have to be confirmed by a court within three days with a 'hygiene emergency prohibition order' – are now contained in The Food Hygiene (England) Regulations 2006. I won't give you chapter and verse here because this isn't meant to be a text book. If you really want to, you can look it up for yourself.

But why do the proprietors still have to go to court if the EHO has already made the decision to close the premises?

A very good question and I'm glad you asked it.

Basically it's to stop corruption by preventing a protection racket.

If EHOs were not answerable to anyone, in this case the courts, they could quite easily threaten proprietors with closure unless they 'paid up'. They could become the Al – Alan or Alana – Capone of the food hygiene world. Or more likely the Halal Capone of it!

Because environmental health departments didn't have delegated powers at the time of this particular incident, Pat had to report to the Environmental Health Committee to ask for permission to take the necessary action. Or in this case, because the committee wasn't due to meet for a couple weeks, she asked the chair of the committee.

This was when the shit really hit the fan. Fortunately it wasn't switched on.

Councillor Cagney was not only the chair of the Environmental Health committee; he was also the Managing Director of the factory whose canteen we'd just inspected.

Oooops!

So to save any embarrassment, and the inevitable crippling publicity, he decided to close the canteen voluntarily – which is always an option.

Now this doesn't mean that if you see a sign in a window that says the premises are 'closed for refurbishment', that they have closed voluntarily to prevent legal action, and to clean the place up. They may genuinely be 'closed for refurbishment'.

Chapter Five

After a couple of months in the job (that's IN, not on – #sore) I was sent to the local abattoir. Not as a punishment you understand, but as part of my ongoing training. Well at least that's what they told me.

EHOs used to have responsibility for inspecting slaughtered animals to make sure that they were fit for human consumption. Unfortunately the function was taken over by the Meat Hygiene Service (MHS). I personally disagreed with the decision because I believe it's important for EHOs to see food from the farm to the fork. But what the fork do I know?

I must admit that this was the day I was dreading. I was convinced that I'd be a devout vegetarian by the end of it. However, and I'm not ashamed to admit it, I actually enjoyed the experience immensely. And I don't mean in a sick (original definition, not the modern one) – Dexter Morgan – macabre kind of way. There was a huge sense of satisfaction at the end of the day that I'd completed the task in hand and actually achieved something. A lot of environmental health work takes an inordinate amount of time to achieve any results and so I would regularly leave work with balls still up in the air (stop it). In contrast, at the end of every day in the abattoir all the dead animals had been inspected and I felt safe in the knowledge that the public could sleep soundly in their beds as their health had been protected because I'd condemned any unfit meat.

Is it a bird? Is it a plane? No, it's 'Meat Inspection Man'.

The best dressed 'superheroes' that season were wearing a cheeky little outfit which comprised of a white cotton boiler suit with press stud fastenings and no pockets, a white adjustable hard hat over a nylon hairnet, a plastic scabbard (and that's not Ron Weasley's pet rat either) on a plastic belt containing a (in my case very blunt) knife and a steel – to sharpen said blunt knife – around the waist. And finally, to complete the ensemble, white steel toe capped wellington boots. It was all very chic – although more Hack it than Hackett, it has to be said.

I must have looked like a missing member of The Village People.

Later I was also issued with a chain mail glove to help prevent me from cutting my hand and so it had more of a Village People meets Michael Jackson vibe to it.

Alan Wilson was the resident meat inspector at the Salchester abattoir and stood out from the crowd, in more ways than one. Not only was he at least 6 feet tall, in his stocking feet (why do we use that term when he wore socks?), very slim with a floppy mop of carrot red hair – okay, it was ginger (he even drove a bright orange Vauxhall Viva) – he was also a perfect gentleman, in every conceivable sense of the word. He cared more about the welfare of others, including the animals that were to be slaughtered, than he did about his own, and in my first few nervous days with him took me under his wing, protected me and nurtured me.

I learnt shed loads from Alan. In fact you could say I learnt an aircraft hanger's worth of stuff from him. I also had tremendous fun with him during our breaks, all of which I will never be able to thank him enough for.

Although I liked meat inspection, the one thing I didn't like was observing ritual slaughter.

There was a large Jewish population in the more salubrious areas of Salchester and their meat had to be slaughtered in accordance with Shechita, the Jewish ritual slaughter according to Halakha. For this the cattle were placed in the casting pen, which held them in place so that they couldn't move, and were then tipped completely upside down to expose their neck. In that position the animal couldn't lift its head because it doesn't have a ligamentum nuchae (I told you I knew what it was) which then makes it much easier to cut the throat. This act has to be performed by drawing a very sharp knife across the animal's throat, only making a single incision, so that the blood is allowed to drain out. The cut has to be performed without pausing, pressing, digging, slipping or tearing and if it isn't then the animal is not accepted for Jewish consumption and is given to the gentiles.

The look of sheer terror in the animals' eyes as they were being upended is one that I will never forget, even though I've tried – hard – many times.

Because the Jews porge their meat, (take out the veins and sinews), they only actually consume the forequarters of the animal, as the rear is too difficult to porge, and so they don't even get the best cuts. They don't eat the kidneys or the intestines either.

Just after animals are slaughtered there is still a certain amount of twitching going on – and I don't mean bird watching, I mean muscle fasciculation. So you can imagine my surprise (too mild – I think complete shock would be more appropriate) the first time I had my hand inside a cow's mouth examining the internal cheek muscles and its tongue licked my arm. You can, I know you can.

Now I like a bit of tongue action as much as the next person, but that was neither the time nor the place for it and not exactly the type of tongue action I would usually have in mind.

I wanted to run, but I was riveted to the spot.

I wanted to shriek in my best girlie voice, 'It's still alive, it's still alive!' but when I opened my, quite considerable, gob nothing came out at all. Not a squeak. Not a peep. So hopefully if anyone was watching me at the time they'd have thought that I was bored and just yawning.

In between that day and my final meat inspection exam I went to many different slaughterhouses in many different authorities and saw and learnt many different things about many different diseases and conditions.

I once, and fortunately it was only once, saw a cow slaughtered that when it was eviscerated it exposed a distended uterus – an indicator that it was pregnant. When the uterus was opened up there, fast asleep, was a perfectly formed calf wearing its 'golden slippers'. ('Golden slippers' is the term used for hooves that have never been trodden on because they are a yellow/golden colour). I couldn't understand why anyone would send a pregnant cow for slaughter? It just didn't make sense.

One of the many different authorities that I continued to learn meat inspection at had three pretty (actually they were ugly) small slaughter houses. The noticeable difference between those and some of the others I visited was the quality of the animals that were sent for slaughter. They were not on a par (they were at least twenty over) with those sent to the others. But that was no bad thing, for me anyway, because I learnt considerably more because of the number and variety of conditions and diseases I saw.

The two meat inspectors that worked there were extremely patient and incredibly helpful and also taught me a lot, as well as compounding what Alan had initially taught me. However, they did like to play the odd practical joke on me. The best one was when they gave me a bull's pizzle – or Willy (or won't he) – turned it inside out and then asked me to identify it. When it's turned inside out it looks like an oesophagus (gullet), which is what I said it was. And believe me that is not a cock and bull story.

Another time they asked me to retrieve a sheep's liver from the condemned area – which was just a cubby hole where the unfit stuff was chucked in – so that I could see what liver fluke looked like. As I tentatively entered the 'hole' I trod on something unpleasantly squelchy. It must have had a huge festering abscess in or on it because suddenly there was an almighty bursting sound as it exuded its pussy (no comment) contents all over my wellies. It reminded me of that well known pantomime – Cinderella. Oh, okay then; Puss in Boots – or in that case; Boots in Puss. Oh yes it was!

In our final year at college the students went to Manford abattoir every Friday afternoon to prepare for the practical meat inspection exam. I'd already been there as part of my '1979 World Tour of Lancashire Abattoirs'.

On one particular occasion, as we strolled onto the slaughterhouse floor, we were greeted by a crowd of people racing hysterically towards us. Even though the situation wasn't in the least bit funny.

"Get out of the way."

"Move it."

"Run for your life." Okay, I made that one up to create a bit of dramatic tension.

"What's the matter, what's happened?" I enquired completely bamboozled. In fact by bam had never been so boozled.

"A boar that they thought was stunned isn't, and when it was dropped out of the casting pen it got up. Now it's on the rampage."

Not surprising really. I mean, being shot in the head with a captive bolt must give you one hell of a nasty headache.

Not only was the abattoir fairly close to the college it was also a large export abattoir and therefore had more sophisticated facilities than the others I'd visited. In fact the pigs weren't electrically stunned there – like the sheep were – before slaughter, they were gassed. However, on the day were taken to witness this marvel of modern technology all didn't go exactly to plan.

The pigs were released from their lairage, nothing unusual in that, but then they started to fight to get into the cages that would take them down into the gas chamber. Now correct me if I'm wrong, but that does seem a tad unusual – actually fighting to be killed first. They obviously didn't have the slightest clue what was about to befall them. They were like lambs to the slaughter – except for the fact they were pigs – and seemed as happy as pigs in shit once ensconced in the cages. They were all making their usual snuffling and snorting sounds and they descended slowly into the chamber, but because the gas level was not at the correct level the noise just metamorphosed into snoring because they were not completely knocked out. I realise it sounds funny (peculiar) that I found it funny (ha ha), but I did. In the end, when they came back up out of the chamber they had to be electrically stunned before they could be slaughtered.

Also in our final year every Thursday morning was spent at the local wholesale market looking at fruit, vegetables and fish, but not necessarily in that order.

We would practice identifying the different food items and, in the case of the fish, inspect them for fitness.

Naturally after handling the fish we would smell – well fishy really – and on the way back to college would regularly stop off for a coffee. However, before we did that we would call into the local large department store and make a bee line for the perfume counter to freshen up. I'm not sure what the mixture of Haddock and Havoc created. Would it have been Haddoc? Or perhaps it was Havock? Anyway after a couple of weeks of this the staff cottoned on to what we were doing and removed the perfume/after shave testers from the counter when they saw us entering the shop. Miserable lot.

We were also keen enough to spend our Wednesday afternoons – which were free for sporting activities – in the local Asian shops looking at, buying and tasting a plethora of – then – exotic fruit. The best way to remember the names of foods is to taste them. I can always remember the first time I tried Guava, I thought it smelt like cat pee, but fortunately didn't taste like it. Not that I've ever consumed cat pee – to my knowledge anyway.

Chapter Six

So, finally, we come to the culmination of my three years of education and training with... fanfare (for the common man), please... the finals!

These exams were split into three parts: the written papers, the premises inspection and the professional interview and food inspection.

The written exams consisted of five three hour papers which were sat every morning of the exam week.

The night before my first exam I decided, after minimal deliberation it has to be said, that it would be wise not to visit the pub this time round – once bitten, twice shy.

Instead, after a mug of steaming hot, hot chocolate – I've been told that I have an asbestos mouth – I went to bed nice and early and refused to look at any of my revision notes. My philosophy was: 'if I don't know it now then I'll never know it.' Not only that, I wasn't one of those students who could cram the night before anyway. It would have only served to confuse the hell out of me and it didn't, and still doesn't for that matter, take much to do that.

The following morning I awoke to a gloriously bright and sunny day – which reflected my mood perfectly. I felt refreshed, confident and ready to face the day and show the Chartered Institute of Environmental Health (CIEH) what I was made of.

Throughout my, fairly considerable, educational life I've found that all examination halls have very similar characteristics. They are usually large, generally austere, decidedly unfriendly, frequently echoey and most definitely always reeking of that unmistakeable 'smell of fear'.

Or maybe that was just my mum's fabric conditioner.

I, obviously, turned up in good time and well prepared with: a pen, a spare pen, a pencil, a spare pencil, a pencil sharpener, a spare pencil sharpener, an eraser, a spare eraser, a selection of sugary delights and, of course, the obligatory lucky gonk.

Well, I had everything except the latter.

Oh alright then, I had the gonk as well. But I wasn't the only one – honest.

The more observant of you will have noticed that I didn't take a drink with me. I didn't know if alcohol was even allowed in exam rooms. This was mainly because I didn't want to have to continually keep getting up for a wee and I certainly wasn't going to insert a catheter for the occasion. I mean that would just have been taking the piss! Not only that, but the invigilator might have thought that I was cheating and had hidden a text book in a plastic bag in the toilet cistern. Also being the clumsy oaf that I am, I would have probably spilt the drink all over my exam paper anyway.

There were lots of other students sitting exams in subjects other than environmental health that week, but we all looked exactly the same – like frightened rabbits caught in car headlights. Not a good look on anyone – including the rabbits.

The first paper was an absolute stinker – and I mean that in the Tom Brown's school day's tradition of the word. Although typically 'Spotty Pemberton' thought it was absolutely spiffing. But inevitably there's always one!

With those fateful words ringing in my ears, "Turn your papers over… now", I frantically scanned the questions to see how many I could answer.

My heart sank.

No, lower than that.

Even lower than that actually!

The confidence I'd had on my arrival quickly evaporated – like a liquid being heated then turned into a vapour and vanishing into thin air – and I was left feeling grumpy, as none of the other seven dwarves were available at the time.

I had to answer five questions out of the eight in front of me, and looking at the paper I was going to struggle to answer even four.

I just stared into space – my final frontier.

What was I going to do?

Faint was my initial thought.

Feign a heart attack perhaps.

I'd trodden the boards, albeit in an amateur fashion, in the past and was sure I could get away with it. In fact I'd previously had rave reviews for my Bottom. They said my acting wasn't that bad either. I did even consider becoming a professional thespian for a while when I was at school, but I didn't really fancy having the operation.

Eventually I managed to pull myself out of my reverie and proceeded to attempt, and I do mean that literally, the four questions that I thought I knew the most about.

That took me little over an hour. Yep, on average, just fifteen minutes each. Boy did I know my stuff!

I now had the unenviable task of deciding which one of the remaining questions I was going to have a stab at.

Or would I have been better to stab myself? – 'Is this a dagger I see before me,' and all that.

Yes, it really had turned into a tragedy.

In pure desperation I decided to play eeny, meeny, miny, moe. Not particularly scientific I know, but it was the best I could do under the circumstances. Besides, I thought that tossing a coin would've attracted too much attention.

Seriously though, I think I'd have had a better chance answering one of the non-environmental health papers being sat by some of the other students. It was that bad.

Not wanting to prolong my agony – I'd had less painful toothaches – I composed a brief message to the examiners that told just a teeny weenie lie, because as we all know, they're nowhere near as bad as a dirty big one. In fact it wasn't really a lie at all; I was just being economical with the truth. I wrote; 'Due to the shortage of time I will answer this question in note form.' I then proceeded to list a sequence of bullet points that I thought, but more importantly hoped, were relevant to the question and then left the exam hall after only ninety minutes.

My bottom lip started to tremble. But my upper lip remained perfectly stiff.

By now the weather had changed considerably. It was starting to turn dark as storm clouds began to gather, which, for the second time in the same day, reflected my mood perfectly.

I was never one for staying behind after exams for the customary 'post mortem'. In that particular case, not only was the exam dead, it was well and truly buried as far as I was concerned. There was nothing else to do except go home and try to get my head together for the next and subsequent days' exams. One thing was for sure, if they were going to be anything like the one I'd just sat, I was going to fail – and pretty miserably too.

The following morning when I met my fellow students, it was clear that everyone had a similar opinion on the previous day's paper and many had left not long after I did.

What a relief. It wasn't just me.

If everyone had done badly perhaps the examiners would realise that there was a Major (or more appropriately a Lieutenant General as far as I was concerned) problem with that paper and be lenient with their marking.

Well we could live in Hope, which, as you are no doubt aware, is just up the road from Charity, where Faith lives – lovely girl.

The next exam was on meat theory which included parasitology and, in the time-honoured tradition, before we went into the exam hall, everyone was asking each other what they'd been revising.

Fortunately for one of my mates, some bright spark just happened to mention the life cycle of the liver fluke (*Fasciola hepatica*) – as you do.

You'd be surprised how easy it is to drop it into a conversation. Try it next time you're out with friends.

Anyhow, he went into a complete panic, "Cr... cr... crumbs," he stammered, even though he didn't have one, "I... I... I've not l... l... looked at th... that at all."

"Not to worry," I reassured him – yes, me of all people – "It's easy. All you have to remember is the mnemonic; **M**anford and **S**alchester **R**owing **C**lub, which stands for; *Miracidium, Sporocyst, Redia* and *Cercaria*." (That's really confused spell-check, it wanted to replace Sporocyst with spookiest). And then I explained what each of them was.

When we turned the paper over what should be the first question on it?

Correct.

You're getting the hang of this.

Well my mate turned to me, with the cheesiest of grins rapidly spreading across his face that eventually stretched from 'ear to there' and would have made the Cheshire Cat look miserable, winked and then proceeded to answer it, as did I.

The rest of the papers went pretty much to plan. After the catastrophe of the first one I was feeling much better about myself and the chances of becoming a fully-fledged – but not fitted with feathers – EHO. In fact the last paper was easy-peasy, simple-pimple and a fantastic way to finish the week.

With the written papers out of the way we were back at college the following week ready, willing and able – well two out of three wasn't bad

– to undertake the practical inspection. This involved inspecting a premise, recognising all the defects, writing a report on our findings and then making suitable and appropriate recommendations. To make it fair, all students had to inspect the same or similar premises.

It seemed strangely unnerving to see everyone dressed so smartly for the occasion instead of wearing the usual denim uniform we had so religiously worn to college for the last three years. We obviously had to look professional on that of all days. We couldn't turn up looking our usual scruffy selves and expect to be taken seriously. For all wants and purposes it was the 'real deal'.

I don't know about the premises being potentially unfit, but the rickety old coach – or 'charabanc' if you want to go all continental – that collected us that day should have been condemned. It was a real bone shaker. Ideal if you wanted to make a Martini for James Bond, but not for transporting a group of uptight students some of whom looked primed to evacuate their bowels at the slightest provocation. I was genuinely concerned that it wouldn't make it to our final destination, but would actually become it.

It trundled slowly through the busy streets of Salchester – that were littered with a variety of junk food rappers; 'M&M' and 'Hot Dog' to name but two – for thirty minutes or so until it came to an abrupt halt outside a couple of rows of dilapidated looking terraced houses.

Well at least the brakes worked!

It was blatantly obvious that the houses were going to be unfit for human habitation because if they weren't there would be very little to look at and even less to write about. That didn't mean to say that we could just get away with declaring them unfit, we still had to justify our decision by identifying most, but preferably all, of the defects.

Here comes the science bit:

The Housing Act 1957 section 4 states that the matters to be taken into account in determining whether a house is unfit are:

(1) In determining for any of the purposes of this Act whether a house is unfit for human habitation, regard shall be had to its condition in respect of the following matters, that is to say:
(a) repair,
(b) stability,
(c) freedom from damp,
(d) natural lighting,

(e) ventilation,

(f) water supply,

(g) drainage and sanitary conveniences,

(h) facilities for storage, preparation and cooking of food and disposal of waste water; and the house shall be deemed to be unfit for human habitation if and only if it is so far defective in one or more of the said matters that it is not reasonably suitable for occupation in that condition.

This legislation has now been superseded by the Housing Act 2004 and the Housing Health and Safety Rating System (HHSRS) which looks at the subject of unfitness in a completely different way – and more comprehensively I might add. Fitness now included 29 hazards such as; lead, radiation and falls associated with stairs and steps. That's without the aid of alcohol and not as a result of listening to the BRIT award winning dance-pop group.

We were split into groups of three to carry out the inspection. The houses were only two up, two down and wouldn't have been able to accommodate a larger group comfortably. Not only that, it would've been inconsiderate to have more than three students invading their homes when the occupants had so kindly given permission for them to be used in such a way.

After we'd been serenaded by a musical doorbell that played the National Anthem, my group were greeted by an adorable little old lady who I immediately wanted to call 'Gran'. She had an abundance of pink rinsed candy floss hair held in place by kirby grips and a smearing of bright pink lipstick across her thin lips. She'd obviously made a huge effort with her appearance for our visit and was wearing, what used to be considered the Women's Institute's uniform, a 'twinset with pearls'. Or in her case a twinset with plastic beads – all in pink. It must have been her Sunday best outfit.

But then why wouldn't she make such an effort? She was of the generation that dressed up when they were receiving visitors.

In a pillow soft voice she introduced herself as Bronwyn.

"Come in, come in," she proffered in her distinctive Welsh brogue. Then with a warm smile and the cool scent of lavender – probably Yardley talcum powder, probably a Christmas present – she ushered us in, but without using a torch.

"Would you all like some tea?" she seemed to sing in that beautiful lyrical accent. "And I've bought a cake as well."

61

Ahhhhhhhhh, I hear you sigh.

Yep, she really was that adorable, which made it all the harder to refuse her generosity because we had to get on with the job in hand – the inspection.

"When you've finished, what you have to do then?" she enquired with a distinct look of hurt in her watery, steely grey eyes.

Feeling somewhat guilty now, we all agreed that if we did have time at the end of the inspection we'd love to join her for tea and cake.

The expression of sheer relief that radiated across her softly wrinkled face was a joy to behold.

At first glance I thought that there was nothing particularly wrong with the house, except for the colour that was. The whole place was pink. Not just the wallpaper and paintwork, but the fabrics too. I'm sure Barbara Cartland would have been jealous. I think 'intense pink' would have best described the colour. It was so intense it was actually hurting my eyes and I thought I was starting with a migraine.

Who shouted, 'drama queen'?

Unfortunately the colour of a property wasn't included in the fitness standard otherwise I'd have declared it unfit for that reason and that reason alone.

So then, what was wrong with the place?

Was it a trick?

Had we been given a house that was actually fit?

I very much doubted it.

It was then that I became acutely aware of an unmistakeable musty odour that usually heralded a problem with dampness.

On closer investigation the defects started to reveal themselves. Most of them had been hidden by her paraphernalia – and I don't mean that in the Sherlock Holmes sense of the word either. What I do mean is the one thousand piece jigsaw (don't call me Rain Man) of the Queen that had been stuck onto a board and was hanging on the wall. "It hides a nasty stain that's lying there," Bronwyn told us. The stain was the result of penetrating dampness. Also there was the glass display cabinet that was fit to burst with a selection of royal family commemorative china – that was probably made there as well – which was hiding the worse case of rising damp I'd ever seen in my, so far, short career.

She dutifully followed us around the house – like a faithful puppy – but not in a suspicious protecting her belongings way, but in a helpful pointing out the defects way, which was really useful.

Thanks Bronwyn.

So, with some time to spare at the end of the inspection, and to her obvious delight, we had time for refreshments.

She brought out what I assumed was her very best crockery: teapot, cups and saucers, sugar bowl and milk jug, as well as side plates all embellished with a thistle motif – not very Welsh. But the piece de resistance was the cake stand, complete with paper doily and a Victoria sponge taking pride of place on it.

I thought I was having tea at the Ritz.

She offered to be mother, poured the tea and then proceeded to sip hers elegantly with her pinkie finger suitably extended – very demure.

"Thank you so much for coming to see me," she faltered as we were leaving, clearly upset. "I've really enjoyed meeting you all. You can come and visit me again anytime you like."

I'm convinced that I saw the merest hint of a tear welling up in her eye as she enthusiastically waved us goodbye, all the time clutching onto her pristine white lace handkerchief as if her life depended on it.

We returned back to the coach where we met all the others and everyone clambered back on clearly in buoyant mood. We weren't quite singing 'ging gang goolie', mainly because we didn't know the words, but someone did start to hum.

Or was that just the smell of stale nervous sweat?

On our return to college we spent the next hour writing our reports all happy in the knowledge that we'd done our best listing the defects and making suitable and appropriate recommendations.

With what seemed like not enough time to catch our breath, the practical exams were upon us. These consisted of three sessions, a professional interview with a panel of, so called, experts; meat inspection and other foods inspection.

Due to the large number of students in my year the exams were held over two days and I was to be examined on the second day.

On the evening of the first day some of my friends visited me at home to tell me how their exams had gone, what questions they'd been asked and what was there to inspect.

They told me that there was a tuberculosis (TB) carcass in the meat exam. But would there be the next day? – TB or not TB that was the question. I was convinced that it wouldn't be, and I was right – for a change!

They also told me about the Blue Ling fish, the bones of which are not white but a blue/grey colour. I'd heard of the fish but didn't know about the bones, but again I was sure that by the time I got there the Blue Ling would have gone AWOL.

The following day I turned up feeling quietly confident only to have it knocked out of me, yet again, at the first hurdle.

My first session was the professional interview with a panel of four examiners that included a doctor – of medicine not philosophy – and a chief environmental health officer, all sat behind the customary 'large imposing' desk.

The doctor was a funny – but not in the ha, ha, sense of the word – little man. He had medium length brown corkscrew hair that looked like coiled springs sprouting out of his head. He was definitely long overdue an encounter with a barber. His bushy eyebrows peeked above his round horn rimmed glasses, like a drag queen's false eye lashes, and matched the thick tufts of hair sprouting from his nasal and aural passages. He wore a well-worn brown elephant corduroy jacket complete with leather arm patches and a breast pocket full of pens. Around his neck was what looked like an old school tie – well it had some kind of crest, a coat of arms or an emblem of some sort on it – that was a bit squewift to say the least. He was just an eye twitch away from looking like a mad professor.

I didn't recognise the Chief EHO even though he was from a nearby local authority. But then again why would I, I hardly socialised in senior management circles. He was a weasely little man with thin mousey-coloured hair and sported the popular Bobby Charlton comb over style. He spent most of his time, when not asking me questions, ferreting about in his briefcase. I just hoped that he wasn't going to badger me too much.

I can't think of any more cute furry animal metaphors.

He also had very harsh, angular features. He actually looked like a Gerald Scarfe caricature of himself and to top it all he had a lazy eye. If it hadn't been for the fact it was just me sat in front of him I wouldn't have known who he was looking at.

I never did find out which council he came from.

Initially I was asked a variety of quite short and fairly easy questions by the doctor and chief officer. I wasn't really sure why the other two examiners were there to be honest because their persistent silence was bordering on the deafening.

I presumed that this easy question tactic was a ploy to put me at my ease before they eventually went for the jugular.

Following the 'easy' questions, I was handed a photograph showing a newel post – the end of a banister rail on a staircase – in what was obviously an old house. Halfway up the post there were some teeth marks.

"Describe the scene and explain what the public health concerns would be," invited the doctor.

I pointed out that the teeth marks were too high up the post to be those of a rodent, unless of course it was on stilts.

No, I didn't really say the last bit, although I was sorely tempted to. The reason I didn't was because the panel all looked as though they'd had a personality by-pass and didn't have a sense of humour between them.

"Going by the size and shape of the teeth marks, they obviously belong to a child," I informed him. "This is an eating disorder known as pica, the persistent and compulsive craving for non-food items. My main concern would be that the property looks old and the paint on the newel post, more than likely, would be old too and therefore probably contain lead, which is extremely toxic, and could affect the child's health in the long term."

It was now that I was, in my humble opinion, quite rudely and certainly unnecessarily interrupted by the doctor. As I was talking he slowly took off his glasses, folded them carefully and then placed them very precisely in their case. He sighed, rather over dramatically it has to be said, and then reluctantly declared, "You can stop right there, you obviously know what you're talking about, so let's move on."

Ah come on, at least let me finish what I've got to say first. I thought you were here to find out what I know, not what I don't know.

I was seriously questioning his parentage.

It was now that they went for the jugular – big time.

"Let's talk about coin operated dry cleaning machines," the chief officer piped up.

Nah let's not, I thought, *because I know diddly squat about them.*

In point of fact that's not strictly true. I knew they were machines that dry cleaned and they were coin operated. Oh and yes, they needed the correct money too because they didn't give change.

It was then that I started to squirm in my, now grossly uncomfortable, hard plastic chair. Beads of sweat were starting to bubble up on my furrowed brow, across my top lip and down the back of my neck. It must have looked as though I was suffering from some, non-contagious, tropical fever or had just eaten an exceptionally spicy vindaloo. I was also starting to sweat in some other places too personal to mention in polite company – which I consider you to be.

Then the inevitable happened. I was shifting in the chair – as though my haemorrhoids were playing up (which I do now have – too much information I know) – when I produced what sounded like an almighty wet fart.

I was dying to laugh.

Why is it that the most stupid things always seem so much funnier at the least appropriate times? Simple things amusing simple minds (not the band) I suppose.

Well, the examiners looked furtively at each other and then accusingly at me.

Come on, I wanted to say, *it evidently wasn't a real one, there's no obnoxious smell attached to it for goodness sake.*

But some things are just better left unsaid.

At the end of the interview the doctor pointed out, quite viciously I thought, "You've said that you didn't know the answer to quite a few of the questions we've asked."

Then why did you continue to ask me questions I didn't know the answer to then? I wanted to protest. But again I thought it prudent not to. In any case, I didn't think it was actually that many.

"What would you do if you found yourself in a similar situation as a newly qualified officer out on the district?" he continued.

"I'd be honest and admit that I didn't know the answer but I'd go back to the office, find out the answer and then get back to them as soon as possible," was the obvious and my truthful answer.

His immediate response was, "Really?" and moreover, in an extremely condescending tone too.

But that had to be the correct answer surely? I couldn't just make something up and hope for the best in case I made the situation even worse.

In the nick of time the bell that indicated the end of the session rang and I left the room feeling completely demoralised and totally baffled. However, even though I felt totally despondent, I was determined not to let it show. I was not going to let them get the better of me and I walked out with my head held high.

Fortunately I had the other foods exam next i.e. non-meat food items. That was my favourite subject and held no fear for me at all. Hopefully it would give me a much needed boost to my confidence.

I'm ashamed to say that I can't describe that examiner to you because I was concentrating so hard on the food that I hardly looked at him at all. Yes, I remember that it was a he and he was very pleasant, but other than that I'm at a complete loss.

Just shows you how reliable I'd be at a police identity parade – in fact I can only just about remember what Sting looks like! I can, however, still describe John Dory, Victoria Plum and Charlotte Potato, although Percy Mon and C Bass are not so easy these days.

There were two ginormous tables in the exam room. One was completely covered with fish and shellfish and the other crammed with fruit and vegetables.

Initially I was asked to take a look at the fruit and vegetables.

"Name any that you know," was the examiner's unexpected, but very welcome, first question. I'd been expecting something a little more specific to be honest.

I systematically ploughed through all of them and rattled their names off one after the other. Guava, pitaya, physalis (no it's not a sexually transmitted disease – it's a fruit), I knew the lot.

Exactly the same happened with the fish and shellfish.

At this point the examiner seemed a little nonplussed that I'd completed the task so quickly.

"Inspect that fish for me and describe what you would be looking for to determine its fitness."

I picked up the fish in question – a cod if you were wondering – and proceeded to squeeze the blood from the gills and look at the eyes etc. Eventually I declared the fish to be fit for human consumption, which was in fact true and not a load of old codswallop – titter ye not.

Then to my surprise, but sheer delight, the examiner asked me about the bones of the Blue Ling. I hesitated momentarily, as if I was delving into the depths of my knowledge for the answer, then tentatively I told him, with a suitably perplexed expression on my face and a questioning

67

tone in my voice, "I think the bones are not actually white but a sort of blue/grey colour."

He looked completely astonished and told me that I was the first student to get that question correct.

I couldn't quite work out exactly who was bluffing who!

I should have taken more time naming the fruit, veg and fish. I should have picked each one up and described its salient features before naming it; i.e. this fish has a silvery lateral line above which the colour is a greenish black, as well as a white belly, therefore, it must be a Pollock, and so on and so forth. Because I'd named everything so quickly there was time left for more questions and of all the subjects he could choose from, he chose the canning of food.

Yet another of my pet topics – not!

My last session was meat inspection and the vet who greeted me outside the detention room – for unfit meat, not the alternative to 100 lines – was easily the nicest examiner of the lot.

The first thing I noticed about him was his feet and his red shoes, not ballet pumps, covered with clear plastic bags sealed around the ankles with elastic bands. It looked as though he had toffee apples on the ends of his legs, which would have given a whole new meaning to the phrase, 'to lick someone's boots'. Apart from that he wore a knee length, long sleeved, protective white cotton coat and latex gloves -'nurse the screens' – as well as a nylon hair net.

He had the most honest and trustworthy face made up of large round kind eyes, a small pug nose and a toothy friendly smile. His whole demeanour was relaxed and he just oozed warmth, which put me at my ease straight away; a skill the doctor (Who?) could have done with learning.

Immediately on entering the detention room my eyes were furtively darting round to see if the TB carcase was still there, but as I'd predicted – using my crystal ball (I can see if I'm coming or going) – it wasn't.

Surprise, surprise!

There was, however, a bovine/cow's liver on the table that looked perfectly healthy to me. I was desperately praying that I wouldn't be asked what was wrong with it, because I hadn't the slightest idea.

So guess what happened?

You're one step ahead of me.

"Take a look at that bovine liver."

Gulp.

"You'll notice that there's nothing wrong with it."

Sigh of relief.

"Explain how you'd inspect it."

Doddle.

Get in!

Next there was a porcine/pig carcase that had a red diamond rash on it.

"What's that condition called?" he asked.

"Swine erysipelas," I said confidently.

"Correct, but can you tell me if it's a zoonosis or not?"

"No," I said rather too confidently.

"Are you sure? What I mean is: is it transmissible to man?"

I'd obviously made a right pig's ear of that!

Very punny – I don't think.

Bit of a clue that you've got it wrong when you're asked the question again.

"Yes," I interrupted, and then went on to explain how it was transmitted to man and what it looked like on human skin, proving that I did really know what I was talking about – for a change.

After many more questions we moved into the next room where there was a horse's head, with an extremely swollen jaw, lying on the table.

"Can you tell me what's wrong with the horse's head?" he asked with a wry smile.

"Well, I think it could be one of two things," I replied innocently and then went on to describe Actinomycosis, more commonly known as lumpy jaw, and Osteomyelitis, which is an inflammation of the bone.

"Do you know, you're the only person to mention Actinomycosis. Well done."

What a swot!

In the distance I could hear the bell ringing again. The vet asked me whether it was the bell he'd heard, but I feigned ignorance – which was fairly easy for me to do – and said I hadn't heard anything.

"It doesn't matter either way," he said, "Because I've heard enough. What session have you got next?"

"Meat inspection is my last," I told him, quite relieved at that point.

"In that case," he told me with a knowing grin, "I think you deserve a drink and if I were you I'd go down the pub."

I knew I'd passed my meat inspection.

I was on top of the world, even if I wasn't looking down on creation.

Chapter Seven

On the first day back at the office – after my final exam – I was put into the Food and Health and Safety team and given my own 'patch' and workload.

"But I'm not qualified yet," I protested.

"You're as good as," was the unwelcome response.

"But I won't be until I'm a registered member of The Environmental Health Officers Association" – as it was called back then.

"No one will know that," was the blunt and disinterested reply.

"But I will," I implored. I was starting to sound pathetic. Okay, more pathetic than usual.

"You're nigh on qualified, we're only waiting for a bit of paper." I could tell that my argument wasn't cutting the mustard.

Yeah, but that bit of paper might say; 'failed professional interview'.

I was glad, however, that someone had confidence in me, because at that particular moment in time I certainly didn't.

"What if someone asks for my authorisation?" I continued to protest, "Mine says student EHO."

"Don't worry – no one will ask for it," was not the answer I wanted to hear, but unfortunately it was the one that I knew to be true.

I could tell that my protestations were falling on deaf ears, and not the pollution principal's either.

The first few weeks out on my own, all I seemed to do was look at accumulations of rubbish in some of the scruffier back alleys of Salchester. I then routinely, and robotically, sent memo after memo to the cleansing department to ask them to remove the said accumulations. It would have been much easier to just say 'please clean Salchester'.

Have I really invested three years of blood, sweat and tears to do this? It would have been cheaper to train a chimp.

Other primates are available.

This was not the first – as you are already aware – or the last for that matter, time that I would contemplate my very existence. I mean, why was I doing that shitty job anyway?

The accumulations ranged from vile-smelling plastic bags full of putrefying domestic refuse to filthy, stained – with goodness knows what – mattresses and bits of knackered old sofas. The former was an ideal food source, whilst the latter were perfect nesting sites, for the infamous and seriously enigmatic *Rattus norvegicus,* or common rat to you – there isn't a posh one. Cue sinister music.

I was expected to go rooting through those disgusting bags to see if there was any indication who the rubbish belonged to.

Not on your Nelly!

I had no way of knowing what I would be putting my hands in. We weren't given any protective clothing in those days or even alcohol gel (in case we drank it?) to clean our hands with so there wasn't 'a cat in hell's chance' that I was going to go delving into that detritus to see if there was anything with a name and address on it.

Who do you think I am anyway? Sherlock Holmes? I ejaculated. Although to be fair, Holmes didn't ejaculate anywhere near as often as Watson did. Read the books.

Six weeks later, and two weeks earlier than expected, I received my results – I'd actually passed. I was now a fully qualified EHO with a certificate and everything.

Well, just two certificates actually. One was my Diploma in Environmental Health issued by The Environmental Health Officers Education Board and the other was my certificate of membership issued by the aforementioned Environmental Health Officers Association.

It was during those first few months as a newly qualified EHO that the 'powers that be' decided – in their infinite wisdom – we were to tackle health and safety inspections quite differently. We were now going to target one specific type of premise at a time. But which was it going to be?

'And the winner is'; there was a lot of fumbling as the golden envelope was ceremoniously torn open followed by a painfully long dramatic pause... 'Funeral directors.'

The announcement was greeted with gasps of surprise, tumultuous applause, floods of 'crocodile' tears and incessantly long acceptance speeches. It wasn't so much the 'Oscars', it was more like the 'Cadavers'.

I never found out what the rationale was behind the decision. But 'ours was not to reason why, ours was but to do and die' – an appropriate misquote under the circumstances. Whether there had been a particular problem in the past or the department had received a significant number of complaints – not from the deceased obviously (unless it was through a medium. I'm a large so it couldn't have been through me) – we were never told.

Communication was always a recurring problem throughout my career and the regular and systematic lack of it usually lead to unsolicited rumours flying around which were much harder to scotch than if the truth had been told promptly in the first place. Ring any bells?

I read up on the relevant guidance available, which wasn't that much to be honest, as I'd never visited a funeral director as a student. Although I'm quite sure the department had signed to say I had (see Chapter Twenty Four).

Some smart aleck – his name was actually Alex, but the hard of hearing principal misheard it – pointed out that it was a dying trade anyway, which I must admit I thought that was dead funny. Sad I know.

Now that I was qualified the department had taken on another student so I invited him out with me on the inspection. I knew from past experience how difficult it was sometimes to be taken out by some of the other officers (if you went to the toilet you were lucky if any of them were left in the office on your return).

John Thomas – I know, but I kid you not. What were his parents thinking? Only Thomas as a first name could have made it worse. Anyway, John was a mature student having already attained a degree in Microbiology. However, even though he'd been to university, it was apparent that he'd never attended the University of Life and was in fact quite green – of all colours.

On his first day in the department he arrived dressed totally inappropriately for work in; faded denim jeans; a faded skull and crossbones t-shirt and faded trainers. And another word for faded is: faint.

You'll understand the significance of that statement shortly.

He must have thought that he was back at university, but soon realised he wasn't when he was sent back home to change into something more comfortable, sorry appropriate – *dopius sodus* (to quote the Latin).

He must have had a very sensitive complexion because he always had a shaving rash or razor burn – whichever is the correct term for those raised welts – on his face and neck. I think he would have been better off growing a beard to be honest.

Unless of course he could only grow bum fluff, in which case he could have got a cat to lick it off.

He was also a painfully quiet, perhaps even shy, person and kept himself very much to himself most of the time, only really speaking when he was spoken to. It was all very Dickensian.

I asked John if he'd ever visited a funeral directors before, perhaps to pay his last respects to a dead relative or friend.

"No," he replied – short and to the point.

"Well, this is going to be an education in more ways than one then," I said as though I was doing him some kind of favour.

When we arrived at the funeral parlour, without an appointment – obviously, we introduced ourselves to the owner, and for the first time ever I was asked for my authorisation. Thank goodness it hadn't been a few weeks earlier; I would have definitely had egg on my face – metaphorically speaking of course. I don't even like eggs, so the chances of actually having egg on my face are in fact zilch – unless someone threw one at me that is.

We inspected the whole of the premises (why wouldn't we?) including the embalming area – which, just for information, is no longer carried out in the same way the Egyptians did when they were embalming the Pharaohs. I asked all the relevant questions, to the best of my knowledge anyway, to which I got suitable and satisfactory answers. There was very little wrong with the place to be honest. A few minor tripping hazards in the office and that was about it really.

Finally, we ended up in the chapel of rest where we looked around to see if there were any hazards to the staff or the public, but there weren't.

When we eventually got back into the car I asked John what he thought about the inspection.

"I didn't like the smell in the embalming area," he said and wrinkled his nose to convey his disgust.

"I agree," I affirmed, "Formaldehyde isn't the most pleasant of smells. I can't see it ever catching on as an aftershave or perfume, can you?"

I laughed momentarily at my own joke.

He didn't.

I didn't think it was that bad.

If you think it's funny then please feel free to laugh out loud – unless, of course, you are sneakily reading this in a funeral directors.

"I did like the way they put those wax dummies in the coffins though, so that you knew what they would look like when they had a real dead body in them."

How am I going to break it to him?
Should I tell him the truth, or should I lie?

There was nothing for it. I was going to have to be brutally honest. After all, as we know, honesty is always the best policy.

"They weren't actually wax dummies," I said all matter-of-factly, "They were in fact real dead people."

Complete silence.

He just stared blankly into the middle distance.

Then the colour drained from him like water down a plug hole. If he'd been naked, and thankfully he wasn't, I'm sure I would have seen it emptying rapidly from his feet.

Then he fainted.

Perhaps I should have lied after all.

When someone faints after a shock all it is, is just a simple vasovagal reflex, nothing more, nothing less. The heart rate drops, the blood vessels dilate and the hydraulic power that forces the blood to the brain diminishes. Result: out cold.

Luckily for me, and him for that matter, I knew that there were five points to consider when treating someone who had fainted and I was now going to have to put them into action.

Thank you Doctor.
Don't mention it.

First: catch the victim. There was no need to do that because fortunately he'd fainted in the car.

Second: lay the victim down with their feet higher than their head so that gravity helps the blood drain back to the brain. That was easier said than done because unfortunately he'd fainted in the car. I somehow didn't think he'd appreciate waking up with his feet on the dashboard and his head in the footwell. So I just titled the back rest back as far as it would go and hoped for the best.

Third: check for a pulse. I did, in his wrist. It was all present although not necessarily correct. It was tapping away with a very unusual rhythm. I reckon Fat Boy Slim would have loved to have sampled it.

Fourth: stimulate the victim. Now the definition of stimulate includes the words; 'encourage or arouse'. So I bet you can't guess which one I opted for. I spoke in his ear, brushed his cheek and even patted his hand, which actually makes it sound as though I opted for the latter! In the end I turned the radio on and turned the volume up which eventually did the trick.

Fifth: persuade the victim to lie still for fifteen to twenty minutes. In John's case no persuasion was necessary.

In the end – when he was fully compos mentis – he sat up, opened the car door and then emptied the contents of his stomach all over the pavement. And yes, there were carrots in it!

Well that all went well then. My first time taking a student out and that's the reaction I get.

It could quite easily have been worse though. He could have not opened the car door!

Chapter Eight

The first food hygiene inspection I carried out, as a qualified officer, was at a Chinese takeaway. Not my choice of premises, it was one of those targeted inspections I mentioned earlier.

Takeaways, not just Chinese I hasten to add, had somehow developed a rather dubious reputation for serving up the likes of dog and cat meat in their food. Not dog and cat food, but the actual animal. However, I can assure you that in all my years inspecting I never came across anything that I couldn't identify or that was considered an unacceptable meat for human consumption in the UK.

In the past, when I was a student, I'd seen various officers struggle with the so called 'language barrier' in those types of premises. Many owners would immediately declare that they didn't speak a word English when they found out it was the district EHO calling. Yeah right!

In an attempt to foil this perennial thorny problem, I thought that I would try a new tack. I went into the premises without any of the trappings that announced I was there in an official capacity, i.e. briefcase, files etc. and engaged the proprietor in polite conversation. I discussed such inane topics as the weather or recent sports events before asking them the directions to somewhere local and well know. Once I'd established that they could speak the lingo I introduced myself, went back to the car for my briefcase and their file and then carried out the inspection in the knowledge that they understood most if not everything I was saying.

Sneaky I know, but it was a means to an end and worked every single time.

So, the first takeaway I inspected was rather uninteresting – that was until I came across a storage area that was cram-packed full, from floor to ceiling, with pallets of cat food. Suddenly my interest piqued. Being the naturally suspicious person I am, and because of the aforementioned 'tales', I enquired nervously what they were for.

"The cat, sir," I was politely informed.

Firstly, legally speaking, there shouldn't have been a cat in a food room – for obvious reasons – and secondly the number of tins of food suggested that it was rather a large cat – as in lion or tiger (or even panther or puma) – with a very large appetite.

"But you shouldn't have a cat in here," I explained (is that an explanation or just a statement?) quite calmly considering.

"Oh, I honestly didn't realise that, sir," he said quite innocently.

I was sceptical.

"Why do you have a cat in here anyway? I asked nervously. Not sure if I wanted to hear his answer.

"To stop the mice, sir."

And that was exactly the answer I didn't want to hear.

"You have mice?"

"No because we have a cat, sir," he proudly told me.

I understood his logic completely, but there was no getting away from the fact that he couldn't have a cat in there.

"As I've already told you, you can't have a cat in here. Can you take it home with you?" I enquired, even though he didn't have a choice.

"Yes of course, if that's what you want, right away sir."

I didn't like the way he was being obsequious and continually kowtowing down to me. He may as well have been saying, 'yes sir, no sir, three bags full sir.'

It was all very disconcerting.

"And can you move all this cat food out as well or I might think that you're using it in the food."

"We would never do that, sir," came his wounded reply. He looked absolutely mortified at the mere suggestion. "It will all be gone by this afternoon sir. Please come back later and check that it is all to your satisfaction, sir."

But I did like the way he was keen to comply with my request.

I went back later that afternoon and he was as good as his word and all the cat food had gone. I never actually saw a cat, but I had no doubt in my mind that it wasn't there either.

Every time I walked past his premises after that incident he would always call me in to show me that there was no cat food or cat there.

How easy was that?

I was, however, being lulled into a false sense of security because everything was not always that easy. Not by a long shot.

There was a café on my patch, well there was more than one – obviously, but this particular one, that I am going to tell you about, I had to keep a close eye on or their standards would invariably slide. Well, turn into an avalanche really.

Although the owner resembled a 1940s film star, Lassie, with his shaggy hair style and wet nose – he had a stinking cold – he was either stupid, had selective hearing or was in the first stages of dementia because no matter what I told him during an inspection, each and every time I visited him, he would usually have forgotten it by the time I visited him again. So, in time honoured fashion, I would have to go over it all again, and again, and yet again.

In spite of all that he was a really nice bloke, even if he was in totally the wrong business. Whenever I turned up there was never anyone else in the café and I wasn't always there on the same day or at the same time. I couldn't understand how he actually made any money.

On the first of the many occasions I visited him, we sat down after the inspection so that I could explain everything that was wrong – and there was quite a bit wrong – and not very much right. I won't bore you with a complete schedule of works, but I will tell you about one significant contravention that might get your stomach churning. On the top shelf of his fridge – which incidentally was not working at the correct temperature – was a plate of raw steak which was slowly defrosting and dripping blood into a pan of bolognese sauce that was on the shelf underneath. He was almost certainly only going to re-heat the sauce, and therefore not kill the bacteria in the dripped blood, and then, in all probability, give some unsuspecting customer an unwelcome dose of food poisoning. Needless to say I made him dispose of the sauce.

But before I could actually launch into my carefully phrased constructive critique, he asked if I would like a drink.

"Yes please," I replied. "A coffee would be most welcome. Thanks very much." I didn't think that his kind offer would be misconstrued as a bribe – especially after I'd tasted it.

Off he skipped, and I do mean skipped, as happy as Larry – and how happy Larry was I had no idea – into the kitchen area to the sound of his enormous bunch of keys clanking away as they swung from the belt loop on his, frayed along the bottom, denim jeans.

He then proceeded to prise open, with one of his many keys, and break the hermetic seal on the largest tin of instant coffee I've ever seen in my entire life. It could quite easily have housed a large family of small rodents – although obviously I hoped that it wasn't. It was a brand I'd never heard of before – and fortunately I've never seen since. And it wasn't granules either, it was powder. Or dust, to be more precise.

He scooped out two heaped spoons, which were much bigger than teaspoons that was for sure, of the cack-coloured dust from the tin into a clear Pyrex mug and then doused it liberally with boiling water. Next, he laced it with a generous pouring of evaporated milk – even though he hadn't asked if I wanted it white or not. He then dropped, from an unnecessary height, two gnarled white sugars cubes into the cloudy liquid with an almighty splosh – and I don't even take sugar. Finally he gave it a vigorous stir, clanking the sides of the mug in the process – which I have to admit I found particularly annoying. Eventually he came back to the table and presented me with a mug of 'sludge', which, I hasten to add, was not the brand name of the coffee – even though it would have been an appropriate and accurate one.

"Are you not having one?" I enquired expectantly.

"Not bloody likely," he laughed, "Can't stand the bloody stuff."

I didn't know if he meant coffee in general, or just that particular 'stuff.'

After blowing on it vigorously to cool it, like I'd been taught as a child, I tentatively took the tiniest of sips.

Mmmmm, I know what you mean.

"Do you mind if I smoke though?" he asked politely.

Well he wasn't in the kitchen preparing food so I couldn't really object on hygiene grounds (it was pre smoking ban), but I didn't think it was a particularly good idea given his cold. The problem for me was that I loathed the smell of cigarette smoke. Still do.

"No, go ahead." I said, even though I didn't really mean it.

I didn't realise that he was going to roll his own until he got his tin of tobacco and packet of Rizlas out of his back pocket. With his large, sand paper rough, hands he laid the golden strands of tobacco along the length of the rolling paper and then brought it to his lips, licked it along the length, and then clumsily rolled it into something that resembled a cigarette. Finally he lit the crumpled tube and inhaled the smoke gratefully, held it deep within his lungs for 15 seconds or so, and then exhaled the cocktail of over 5,000 chemicals out through the side of his mouth and then extinguished it in the black Bakelite ash tray on the table.

Was it really worth the effort?

I did my best to ignore the drink throughout our ensuing conversation, but he kept on reminding me that it was there and insisting that I drink it before it got cold.

Hot or cold sludge, does it really matter?

It didn't even have an aroma and actually tasted like mud. Or what I imagined mud to taste like at least. I'd probably eaten some as a child, like we all do, but I couldn't remember the exact flavour it had and I'm quite sure that different mud's must taste differently anyway.

I could actually feel the powder, that hadn't dissolved properly – if at all – filtering out on my front teeth like tea leaves in a strainer. It was incredibly sweet, due to the evaporated milk and sugar, with a nasty bitter after taste, due to the coffee being 'cheap and nasty' – which would have been another appropriate and accurate brand name.

Another cause for concern was that every time I looked at him I was totally mesmerised by a, small but quite noticeable, growth on his face – just under his right eye – that looked just like a miniature cauliflower. If it was a mole he definitely needed to get it checked out by a doctor – and pretty smartish too. I'm sure that I must have been staring at him for overly long periods of time, but because it was so close to his eye he must have thought that I was just making eye contact with him whilst concentrating intensely on his every word. Little did he know!

Each and every time I went to visit him – without fail – from that day on, he would always automatically make me a cup of that, that... whatever it was. I can't actually find the words to describe it, probably because they haven't been invented yet. Perhaps I should create my own: Nasti cafe. It could quite easily have been used to torture foreign spies with and I have no doubt whatsoever that they would 'spill the beans', at the merest hint of its use.

The level of powder in the tin never actually seemed to go down between my visits. Was it a never ending tin of coffee? Or was there only me drinking the 'stuff'? I think in my heart of hearts I knew the answer.

I could see him making it as soon as he clocked me crossing the road on my way to his premises even though I regularly approached from

81

different directions. And he wouldn't let me leave until I'd drunk it all, every last drop – bloody masochist.

Not only that, but it always went through me like a dose of salts and within ten minutes of me leaving his premises I would be in desperate need of the bog (to use the vernacular) and the public toilets around there did in fact resemble a bog the majority of the time.

I eventually made this my reason for not partaking in a beverage when I was there and he reluctantly, but graciously, accepted my excuse – sorry, reason.

I now had a rough idea why the café was always empty. Certainly no one was going in there for a coffee, that was blatantly obvious, and that was also obviously why the catering tin never seemed to empty. I never found out what his tea was like.

"I've just had a tip off that illegal meat is being delivered to the butchers in the High Street," Peter Bowden declared as he burst through the office door (not literally) with his coat over his left arm and his car keys in his right hand. You remember Peter; he was one of the blokes that interviewed me. Although he'd reached the dizzy heights of Deputy Chief Environmental Health Officer he didn't just sit all day in his 'ivory tower' contemplating his navel. In fact there was nothing he hated more than being trapped indoors and was always more than happy to get his hands dirty at the slightest opportunity – and get involved in the real work as well.

I looked around the office to see who he was talking to, but there was only me in there. So it must have been me.

"Come on, let's get down there pronto," and with that he hurtled back through the door and before I could even get my coat on I could see him getting into his car in the rear car park.

The car was actually moving as I clambered into it. We must have looked like Regan and Carter from 'The Sweeney', except for the fact that neither of us looked like John Thaw or Dennis Waterman. If only we'd had one of those blue flashing lights that you stick on the roof of the car, although the way Peter drove you'd have thought we did.

When we arrived at the butcher's shop – within moments of leaving the office it seemed – there was a plain white van outside with what appeared to be several lamb carcasses strewn across its floor.

After Peter had parked the car – well abandoned it at the side of the road really – we took a closer look inside the van. There were in fact four lamb carcasses on the floor and on closer inspection we noticed that none

82

of them had been stamped. (When animals are slaughtered in an approved abattoir they are inspected to make sure that they are fit for human consumption. If they are, then they are stamped by the meat inspector with a stamp that indicates which abattoir the animal was slaughtered in and who inspected it).

"You wait here while I take a look inside the shop," Peter ordered, but not in a bossy way.

"There's three more in there," he informed me when he finally re-emerged. He'd obviously had a really good rummage around. "We're going to have to seize all seven of them."

I'd never done anything like this before. It was really exciting stuff.

"You wait here and make sure the van driver doesn't take off with those carcasses," he said as he pointed to the four in the back of the van. "I'll go and get a J.P. (Justice of the Peace) to come and condemn them."

It may all sound over-dramatic, but we had no way of knowing how the animals had actually been slaughtered or even if they were fit enough to be slaughtered in the first place. If they had been ill before they were killed they could not be used for human consumption.

Okay, so how am I supposed to stop this guy from just driving away? Lay down in front of his vehicle? Not bloody likely.

I told the driver that he couldn't go anywhere – and he didn't. I know, what were the chances? After what seemed like eternity – and a day, Peter arrived back with the J.P. and the meat was formally condemned. We even persuaded the van driver to take it to the incinerator – with me in the van as well to make sure that he did – for it to be disposed of. Now you can't get more accommodating than that, can you?

He also kindly answered all our questions about who he worked for and where he picked the meat up from. It turned out that it was all going on somewhere in another local authority area, and so we 'happily' informed that authority so that they could 'stamp out' the illegal slaughtering.

It turned out that the animals were being slaughtered in lock up garage in the middle of the night.

Chapter Nine

Tell me, why is it that most people think that condensation is dampness?

My first 'solo' complaint of this type was from a woman who claimed that she had penetrating dampness in all four walls of her kitchen. Nowhere else in the property, just the kitchen.

Thinking back as I'm writing this, I don't understand why I was involved in this complaint at all when I was in the Food and Health and Safety team.

That's strange, I said to myself, because no one else was listening – as usual. *At least two, but maybe even three of the walls would be internal depending on the design of the property, so that was, in fact, impossible.*

Mrs Horner lived slap bang in the middle of a long row of red brick, Flemish bond – just showing off – terraced houses. None of them had a front garden, only a front step that paved the way between the pavement and the front door. Mrs Horner's step was heavily coated in 'red lead' (Pb_3O_4), as was the fashion of the day, and was spotlessly clean into the bargain.

On either side of the step were two gnomes that stood like sentinels protecting the residents of the property from – well who knew what! I mean, let's face it, how good would a gnome be in a crisis? Usually, at the first sign of any trouble, they're wandering off – just like a Gnomad.

I'm not being funny though – but you didn't need me to tell you that – I was really surprised that they were still there and hadn't been stolen. She didn't exactly live in the most salubrious of areas.

What am I saying? Who in their right minds would want to steal a gnome for goodness sake, never mind two of them, and two of the ugliest ones you could ever hope to come across for that matter.

Mrs Horner filled the doorway completely when she opened it with an accusatory "Yes?" hissed in my general direction. She was an ample-sized woman – to put it mildly. Her head was encased in a mass of sky blue plastic hair rollers that were held neatly in place with the aid of a pastel blue – see through type material – head scarf, that I'll say was chiffon, but probably wasn't. She wore a drab polyester blouse over which hung a drab, pattern less, calf length cotton pinny. Her feet were encased in faux fur trimmed moccasin style slippers and her hands sheathed in yellow rubber gloves. And finally, to complete her cheeky little outfit – without a word of a lie – she was wearing wrinkled stockings.

I have to admit, I thought she looked a bit of a battle-axe. Well a lot of a battle-axe actually. Not particularly kind I know, but true never-the-less.

She stared at me with naked hostility in her eyes. Her arms were folded tightly across her, somewhat considerable, chest and there was something almost defiant in her stance. All she needed was the rolling pin.

I bet her husband was never late back from the pub!

She also had two strange egg-shaped baubles dangling from each, out of proportion, pendulous ear lobe. Perhaps her husband had been late back from the pub then – but only the once.

"I'm from the environmental health department," I advised her.

She looked at me quizzically, or maybe even slightly confused.

"About the dampness," I continued.

"Oh yes about time come in and I'll show you where it is and don't forget to wipe your feet," she said in one – unpunctuated – sentence, without pausing for breath, and with that she turned on her heels and stomped off down the hall. She sounded like a herd of elephants in a china shop.

I know that's a mixed metaphor and it should be a bull, but please hear me out.

She left a trail of clattering trinkets in her wake.

I stepped very carefully over the front step with a gangly lunging motion that I was positive would have gained me honorary membership of the 'Ministry of Silly Walks'. I certainly wasn't going to be the one to get the step dirty. She wouldn't be getting a second pair of earrings out of me to add to her accoutrements that was for sure! I then obediently wiped my feet on her husband – sorry, the doormat.

The hall had a 'jumble sale' of bric-a-brac – all emblazoned with the name of the destinations the items had been bought from – covering every conceivable flat surface. It included a collection of; miniature brass gongs; porcelain bells and chalk ash trays – to mention just a few categories of the vast array of 'objet d'art' on display. There were some that I hadn't got a clue what they were, and I wasn't going to ask either. They were all just settling back down again as I passed them after 'Nellie' had stormed past.

As I entered the small, probably now referred to as compact and bijou, kitchen I noticed that the walls were not plastered – they weren't even tipsy – but they were thickly coated in battleship grey gloss paint. There was a traditional pulley clothes airer suspended from the ceiling with a couple of 'off white' nylon shirts draped over it that systematically dripped water onto the quarry tiled floor, like some ancient Chinese water torture. Also, nestling in the single drainer stainless steel sink, was an old fashioned, wooden framed, metal ridged, washboard. Perhaps she was a member of the local skiffle group; because she certainly had the jugs as well. And sailing slowly, but quite regally, down the draining board, past the hand operated mangle wringer, on a sudsy lather was a bar of Fairy Mild Green Household Soap.

The whole scene looked as though it belonged in a museum.

Finally, to complete the tableau – sat firmly on the stove – was a catering size pan of water boiling away frantically and producing enough steam to fuel the Flying Scotsman for at least an hour or maybe even two. Unsurprisingly, there were no windows or even the back door open and the condensation pouring down the, two outside, walls reminded me of Niagara Falls – from photos I'd seen, I'd never actually been there. I had, however, been in steam rooms that hadn't had as much steam in them.

I enquired, genuinely interested, what she was cooking so violently.

"Tsch" she admonished. "I'm boiling my knickers," she went on to inform me with a look that said 'stupid boy'.

Not a recipe I was familiar with, or wanted to be for that matter.

Was she boiling them as opposed to coddling or poaching them? I wondered, as I didn't dare ask. I didn't think that her 'funny bone' would be connected to anything.

Perhaps it was from one of those '101 things to do with...' cookbooks.

Did it produce some sort of broth or consommé?

If it had been men's urine stained boxers in the pan would that have made Cock-a-leekie soup?

Just asking.

"It's the only way to get them really clean," she admonished.

I'll take your word for it.

I dreaded to think what she'd been doing to get them so dirty.

She then hoisted a pair out of the pan using her laundry tongs, which looked like a pair of oversized chopsticks, and inspected them thoroughly. They weren't knickers; they were bloomers – and not the kind you get from the bakers either. They could quite easily have been a substitute for a wind sock at the local airport.

I asked her if the walls were always damp to which she replied, "No." She was honest if nothing else.

"Have you ever noticed any particular time when the water runs down the walls like it is now," I asked in a voice laden with sarcasm.

"Yes," she said, "Every time I'm in here doing the washing or the cooking."

"No other time?" I needed her to categorically confirm it.

"No."

I then went on to explain that her problem was in fact condensation, not dampness. I told her that when she was boiling things in the kitchen: vegetables, knickers, but not in the same pan – hopefully – the warm moist air would hit the cooler walls and then turn into condensation. What she needed to do was ventilate the room when she was engaged in that type of activity by either opening a window or the back door.

She looked at me menacingly through her beady little eyes; I could tell that she wasn't convinced.

"I'm not convinced," she snarled spraying spittle into the air.

The audacity of the women!

I bet she could be a right cantankerous old bag if she wanted to be!

To prove my point I asked her if I could use a damp meter to measure the moisture content in the walls, and she readily agreed.

There are two types of damp meter: a small hand held one that only penetrates the plaster and doesn't really show if the brickwork is wet and the larger one for which holes have to be drilled into the brickwork to determine if it is indeed damp.

I drilled two small holes into the walls, placed a probe in each – which were a bit like metal knitting needles – and then measured the current between the two, because water conducts electricity. The higher the reading the more moisture is in the brickwork.

Hers didn't even register. The walls were completely dry.

Told you so!

I left her mumbling something to herself under her breath. I doubted that she would take my advice. I doubted if she would or had ever taken anyone's advice in her entire life. As far as she was concerned there was only one right way and that was her way.

Never mind, you can't win them all. But a few now and again would have been nice.

Chapter Ten

I guess all workplaces have some sort of initiation ritual for their new members of staff and Northford Environmental Health Department was certainly no exception. Thankfully it didn't involve having my head flushed down the lavatory or being paraded around the council offices naked with nothing more than a probe in my hand (cheeky). Although on reflection they may have been marginally preferable to what they did have in store for me.

Because I was the new kid on the block -- not from the American boy band -- and everyone else had already visited this particular complainant, it was now my turn to investigate her latest in a long line -- so long it would have almost stretched to the moon, and back -- of grumbles.

Yes, I was to be thrown in at the deep end, without any buoyancy aids, or even a life guard in attendance for that matter, to see if I would sink or swim.

Fortunately though, I had been given a potted history of the continuing saga of Mrs Hickling so that I didn't arrive totally unprepared. It actually made 'War and Peace' look like a cheap magazine in comparison -- middle shelf, not top.

How kind.

Little did I know that nothing could have really prepared me for this visit!

I was told that she'd been complaining, on and off -- more on than off -- for several years. None of her complaints were rational or within the remit of the Environmental Health department, but no matter how many times she'd been told that nothing could be done for her, she would still complain as regular as clockwork. Or someone on a high fibre diet, if you prefer.

I would now like to, if I may, take this opportunity to give you just a couple of examples of her frequent complaints so that you can see what you think for yourself.

Every autumn she'd collect a selection of dead leaves, from deciduous trees, take them to the department saying that there was something sinister at work that was causing them to drop off the trees that needed to be investigated immediately. She would also regularly send a variety of soil samples, in envelopes through the post, asking for them to be analysed – but never actually saying what for.

I was also alerted to the fact that her husband was an invalid (a horrible and derogatory 1980s term which has now been replaced with disabled). He only had one leg as a result of a road traffic accident and spent the majority of his days parked up on the local hillside in his invalid carriage (now more appropriately called a mobility vehicle) reading vast numbers of library books to steer clear of his wife.

It all boded well.

So armed with this useful background information I headed off to see what Mrs Hickling was complaining about this time.

To my amusement, I also found out that my illustrious colleagues had informed her that they were sending a northern expert (in what?) to see her this time.

What a rotten lot.

I'd only been qualified for eighteen months and besides, I didn't think my accent was that pronounced.

Obviously it was.

Mrs Hickling lived in the picturesque hamlet of Astley Cressett, which consisted of a cul-de-sac with just seven bungalows in it. Not six, not eight, just seven. There wasn't a shop, or a pub, or even a church, just the seven bungalows.

As I pulled up outside her property, everything seemed perfectly normal to me.

What exactly is normal? – discuss.

How wrong could I be!

I had only just grabbed hold of her knocker (behave) when the door violently swung wide open almost dragging me with it. She must have been poised behind it, like a coiled viper, waiting to strike.

When I'd eventually regained my balance I became acutely aware that I was face to face with, what could only be described as; eccentricity personified.

She could have only been in her late forties, but looked the wrong side of sixty. She was wearing a conglomeration of 1960s hippy clothing that Mamma Cass would have been proud of. I mean, where on earth would you get that stuff from? eBay didn't exist back then. Unless of course it was all her original gear, in which case it would've been like 'far out man'. Her hair, which was considerably more salt than pepper, was wildly tangled and gave the distinct impression that she'd recently been 'dragged through a hedge backwards'. She even had flowers in it, real ones, from the hedge I surmised. I thought I'd travelled back through a space-time vortex and had arrived at Woodstock, not the village in Oxfordshire, but the 1969 American festival. And her shoulders were covered in what looked like white powder which I presumed was an accumulation dead skin cells from her scalp, but I could have been wrong.

My jaw duly dropped and hung open in total disbelief. I must have looked completely gormless, i.e. having a distinct lack of gorm.

I barely had time to introduce myself, I could have been anyone for all she knew or cared, before she started ranting on about her latest crisis.

She hadn't listened to a single word I'd said.

Not the first person to do that it has to be said, and I was quite sure that she wouldn't be the last either.

"Come along with me young man," she hollered in a shrill, yet posh, schoolmarmish voice, and then marched round to the side of the bungalow.

Hup two three four!

I did my best to keep up with her.

"What do you think of that?" she snapped with both of her hands resting on her ample hips and her size tens planted firmly apart as if she was a 1950s P.E. mistress. I bet she was wearing navy blue knickers as well.

What do I think of what? I thought, as I wasn't quite sure what I was supposed to be looking at. So rather timidly I enquired to what she was referring.

"The wall, the wall you blithering idiot, isn't it clear?" she spat with actual spittle spraying from her mouth as she did so.

I'd been called an idiot before, many times in fact, but never a blithering one. How frightfully up market!

Well no, it wasn't clear to me at all. Or rather, it was as clear as a crystal glass that had been through the dishwasher – more than once.

The outside wall of the bungalow had been pebble-dashed, with pebbles, as had the other six bungalows. This was nothing unusual, in fact it was a fairly typical design feature of properties of that age and it looked in pretty good condition to me. So what on earth was she wittering on about?

"Can you be a bit more specific please?" I enquired cautiously, at which point – with her impatient eyes raised incredulously towards the heavens and a deep sigh of pure exasperation – she proceeded to be more specific.

Have you ever wished you'd never asked a particular question?

Don't pretend you don't know what I'm talking about. You know precisely what I mean. When you greet a friend or colleague and politely ask how they are only for them to launch into some diatribe about all their ailments and woes. *I didn't really want to know. I was only being courteous and exchanging social niceties. Just say fine thanks, how are you? Then I'll say fine thanks and then you can go away.*

Well, I was about to regret asking her to be more specific. The difference here was that it was my job and I was being paid to ask!

The problem, I was reliably informed, (yeah right) was to do with the black pebbles in the walls. It would appear that they hadn't always been black. Of course they hadn't. What had happened was 'It' had got into some of the other coloured pebbles, come back out, gone back in again (are you following this, because I had difficulty at the time) and that had then turned them black.

Not knowing what to say I just nodded affably, pretended to make copious notes and hoped against hope that I looked as though I knew what I was doing.

Now that would have been a first!

She then promptly turned about face and rushed headlong down the garden beckoning me to follow.

What choice did I have?

In truth I had two choices: I could either have followed her; or I could have just made a run for it.

Can you guess which one I choose?

Oh ye of little faith.

I followed her of course.

I bet you're now thinking that I'd gone completely doollay.

Maybe I had. Mrs Hickling most certainly had, and I don't mean that in a callous or un-politically correct way. She undeniably had some mental health issues, an extremely taboo subject in the early 1980s. Not only was it not really talked about then, but it wasn't treated in the same way as it is today either.

92

Whether her mental state could in some way have been attributed to the Paul Kantner quote; 'if you can remember anything about the sixties, you weren't really there' – if you know what I mean – or a bone fide medical condition, I never found out. But there was a very distinctive 'scent' that followed her wherever she went – and it certainly wasn't 'Opium'.

The garden was very long, approximately 90 feet, or 30 metres if you've been metricised – which I hope wasn't too painful. It was also devoid of any plant life, no weed (not even that type) or even flowerpot men for that matter.

I was now poised for her to tell me that 'It' had decimated the garden as well. But no, she didn't say a dicky bird, and funnily enough there weren't any of those either – and I certainly wasn't about to ask if 'It' had either.

The only thing in the garden was an old fashioned/traditional washing line that stretched the full length and had started to flutter and flap about in a stiffening breeze that suddenly materialised from nowhere. It was tied to two metal posts at either end of the garden and for a prop she used the branch of a tree that had a fork at one end – but no other cutlery. The line was slotted into the fork and the prop was hoisted up so that the washing could be supported for that perfect drying experience.

For those of you who have only known rotary driers, or perhaps just tumble driers, you probably have no idea what I'm rambling on about now. For that I can only apologise, but it is an integral part of the story, so please bear with me. I'm sure if you 'Google' washing line you'll find out what I'm going on about.

'Yahoo' it doesn't sound quite right somehow.

Also, while you're at it, look up 'dolly pegs' and see the multitude of things you can do with them – Blue Peter style.

She then, with a dramatic sweeping gesture, pointed to the aforementioned prop and informed me, rather haughtily, that yesterday the branch had been a fully grown tree and 'It' had done that too.

"Okay," I said, disbelievingly, in a sort of American Deep South drawl.

"Is there anything else?"

Stupid question I know, but it had to be asked, and of course there was.

"This way," she cried, as if leading a group of weary tourists round an uninteresting historical site and off she scurried. All she was missing was a brightly coloured umbrella – and that is not a euphemism.

She then led me reluctantly – me not her – over to the coal bunker at the back of the bungalow. Leaning up against it was an old coal shovel that was so rusty – probably due to many years of neglect being left outside in all weathers – that it had a substantially sized hole in the middle of it. In fact it was more hole than shovel really.

"That was a brand new shovel last week and look what 'It' has done," she said as her hair started to get longer as it untangled in the ever increasing wind.

I was struggling to understand her now because alternate words were being regularly whipped away by the persistent gusts that caused the white cotton wool clouds to race across the sombre sky as if they were late – for a very important date.

"I find that very hard to believe," I bellowed, in case she couldn't hear me. But she did, either that or she could lip read.

"That's how quickly 'It' works," she retorted as the last word of my sentence departed my lips.

I was speechless, which was pretty unusual for me – and you are only allowed to agree with that statement if you know me! And I was starting to get cold too.

"I'm now going to show you 'It'," she suddenly announced – rather proudly - as if she'd discovered a cure for cancer or something else equally life changing.

This I couldn't wait to see.

"If you take a look inside the coal bunker, you'll see 'It'."

Drum roll please.

So, throwing caution to the wind (of which there was now plenty, or should that have been caution I've got wind? – as I was a bit nervous) I opened the door of the bunker, albeit rather tentatively. I had no way of knowing what fate awaited me as she didn't 'give us a clue' – and Una Stubbs wasn't there to give me one either (oh er).

Was there something weird and wonderful lurking in the gloom?

Would 'It' leap out and attack me?

Tune in same time, same place next week to find out.

Previously in Chapter 10 our intrepid hero was embroiled in an incredibly difficult and very complicated complaint that was testing all his powers to the limit. Here's what happened next:

I stared intently into the inky darkness for a few minutes until my eyes adjusted to the light, or rather lack of it. All I could, just about, see was coal and plenty of it. That's plenty of coal and not plenty of 'It'. I still didn't know what 'It' looked like.

Now that may seem perfectly logical to you, but in this particular case, logic had gone on holiday – and it hadn't given a return date.

What was I supposed to be looking at?

Was there something in there other than the coal?

Obviously Mrs Hickling was seeing something I wasn't – even without the aid of a hallucinogen.

Then the penny dropped, so I promptly bent down to pick it up again and put it back in my pocket – the one without the hole in this time – because as we all know; 'if you look after the pennies, the pounds will look after themselves', or so the old adage goes.

At last, I had my long awaited eureka moment – although I didn't actually go running down the street naked shouting 'eureka' you'll be glad to know.

If you recall, when you look at coal in the dark it has a sort of silver sheen to it. Well, when I described this or 'It' to Mrs Hickling – bingo, I'd got a full house. At last we were singing from the same hymn sheet, even if it wasn't a song I was familiar with.

Instantly her whole demeanour changed. She'd suddenly found an ally, someone who finally understood her. Well that obviously wasn't true, but I wasn't about to destroy that delusion for her.

She then calmly invited me into her home to show me what 'It' had done in there.

It was fairly obvious that the bungalow hadn't been maintained at all over the years. The first thing I noticed was some crack(s). The plaster on the walls was obviously the original and was now crazed with hairline and settlement cracks that looked like the veins on a dried out sycamore leaf. Like the ones she'd religiously taken to the department. Unsurprisingly 'It' had made those.

All the furniture was old and tired, I think exhausted would have been a more appropriate term, and hadn't been cared for over the years.

The floor was covered in a scuffed, black and white, chequer board linoleum – or more likely oilcloth. The joints – I'm talking about the ones between the strips of oilcloth – were cracked and curling up and had been clumsily stuck back down with insulating tape. And in the middle of it was a round long tufted rug emblazoned with a hallucinogenic multi coloured swirling pattern. It wasn't so much Lucy in the Sky with Diamonds, more Lucy on the Shag pile with Dandruff – assuming her name was Lucy of course.

The sofa and chairs were strewn with a collection of Indian inspired paisley patterned throws that had prevented the all too evident spillages – of what I wasn't sure – from getting through to the furniture underneath, along with an array of vibrantly coloured, mirror bejewelled, cushions.

The curtains, which hung listlessly at either end of the bay window, must have shrunk the last time they were washed – whenever that was – because they were a couple of inches too short. They also had a few vertical rips in them; from a cat scampering up them by the looks of it – even though there was no other evidence of a feline in residence.

Yes, you've guessed it, all the scratches, scuff marks, tears and stains were the fault of 'It' as well.

'It' certainly got about!

By now I thought I'd seen and heard it all and the situation couldn't get any worse. But I was wrong. She'd saved the piece de resistance for last. 'It' had come one night and dissolved her husband's leg.

Ta da!

What could I say? I was completely flabbergasted. In fact my flabber had never been so gasted.

I had to surreptitiously stifle a snigger before it became a full blown guffaw. Unprofessional I know, but knowing about her husband's accident, her story did strike me as funny – peculiar.

I now needed to make my escape – sooner rather than later preferably. There was very little point in continuing to discuss the problems that only really existed in her head. She was obviously as nutty as squirrel shit. I, therefore, played my ace card. Continuing with the pretence that I was a 'northern expert' I informed her that I was involved in a major national research project that was endeavouring to correlate strange phenomena, such as she was experiencing, with sightings of UFO's.

Absolute genius! Not only did she believe me, but she was totally impressed with me as well. Finally she'd found someone who knew what they were doing. And she was the first and last person to ever think that.

I explained that this research was very complicated and time consuming and because I had all the data I needed from her, there would be no need to contact the department again with any more complaints. So after a considerable amount of umming and ahing (Macmillan dictionary spelling) she reluctantly agreed to my request.

As I was leaving her husband pulled up outside and called me over. He looked considerably younger than Mrs Hickling, even though he did look worn out. It mustn't have been easy living with her, and her paranoia, day in day out. I found it quite exhausting for the short time I was with her. He apologised profusely for having wasted my time and said that it was a psychiatrist she needed to see, not me, but unfortunately she wouldn't go.

The next thing I knew she came hurtling towards us brandishing a copy of, what looked like, the Radio Times and then proceeded to beat her husband over the head with it while screaming, "What are you saying about me?"

During the commotion I discreetly made my exit stage left.

We didn't hear from Mrs Hickling again for several months after that. When she did finally ring the office again the chief was there and answered the call. After listening to her ramble on about another spurious complaint he told her that she was a 'nutter' and that she should leave us alone.

Well that's one approach I suppose.

And she did.

And a successful one as it turned out – even if it wasn't exactly PC.

Chapter Eleven

Greater Winlock was a small, some would say charmingly rustic, medieval market town with a labyrinth of higgledy-piggledy narrow streets to explore. It also boasted a variety of black and white half-timbered buildings, including a few that were Grade II listed, and a plethora of old limestone properties; the majority of which were domestic, the remainder being commercial, to admire.

Well I would notice all of that wouldn't I!

Come to think of it, the description sounds like the introduction to a tourist information brochure on the town – for which I am available to write, for a suitable fee.

It should not, however, – under any circumstance whatsoever – be confused with the even smaller Lesser Winlock.

As if you would.

The town originally grew around the abbey and monastery which were founded around 1015 – just in time for morning coffee – by Merewalh, a son of King Penda of Mercia. The town received its Charter from Edward IV in 1468 granting Borough status and the abbey continued to prosper until the Dissolution of the Monasteries in 1539 – just in time for afternoon tea.

Here endeth the history lesson. If you'd like to take your maths books out and turn to page 34 we'll take a quick look at quadratic equations – but only a quick look, mind you.

I'd received a complaint, via one of our support staff, from a Mr Adams – who lived in Greater Winlock – that the roof to his home was leaking. There were no other details, just his name and address – not even a telephone number.

That's a bit odd (if only it had remained at odd), I thought to myself, because there'd been no rainfall for well over a fortnight. So I was intrigued to find out what he was actually complaining about.

As I rang the doorbell I heard the muffled sound of a single dong somewhere in the depths of the house. Or maybe it was a bong. It was hard to distinguish between the two really as they both sound so similar.

I was just about to ring again when I heard two heavy bolts clunk solidly as they were slammed across the door followed by a key turning tentatively in the lock. Then the door slowly creaked open with a high-pitched screeching noise that sounded as if someone was being murdered. Well, on those tacky television shows anyhow.

It was as though I was starring in some low budget horror film. So low in fact that the actors had to pay to be in it.

Still, it was nothing that a good squirt of WD40 couldn't rectify.

When the door was fully open, I was aware of a very tall (if I'd stood on my tip toes and stretched my neck to its full extent, I'd have still only been looking at his chest), extremely thin (I've seen fatter bean poles) man with a ghostly pallor and deeply sunken dark eyes – his pupils were so large, as they strained to let the light in, they obliterated his irises – accentuated by the dark circles that surrounded them.

He was wearing a dark morning suit, as if he was actually in mourning, and stood bolt upright in the doorway staring down intently at me.

Not in the least bit un-nerving – much.

He looked like an old fashioned undertaker – minus the top hat. In fact he looked more like the corpse to be honest.

I desperately wanted him to say, 'You rang?' because he was the spitting image of Lurch from the Addams family. A coincidence or what?

I couldn't make this stuff up if I tried.

Or was it Nosferatu?

Perhaps he was preparing for an audition for the next Tim Burton film.

Or even simpler – he was anaemic. But no one could be that anaemic surely?

Instead all I got was, "H-e-l-l-o," in a painfully slow and deliberate, adenoidal voice.

"Environmental health," I said in what must have sounded like a fast high pitched squeak in comparison. He must have thought that I was on helium.

"I k-n-o-w." His face looked completely deadpan – actually, you could omit the pan. "E-n-t-e-r."

I was hastily hustled off of the pavement and straight into the living room as there were no halls or porches in properties of that age in that area. No sooner had I crossed the threshold than he furtively closed the door behind me, locked it, bolted it and then dragged a dark heavy velvet curtain across it.

Cosy.

The small sash window, which also opened out onto the street, was also veiled in a similar fashion. The only light available to illuminate, and I use the term loosely; the room was from an antiquated standard lamp – with the smallest wattage bulb known to man – faintly glimmering in the far corner.

Correction, it wasn't cosy; it was depressingly gloomy and alarmingly claustrophobic in a strangely sinister way.

My pupils were probably so large now, as they strained to let the light in, that they must have obliterated my irises too.

No wonder he was so pale. I wondered if he'd ever seen daylight for any longer than just opening the front door.

Did he ever go out?
Could he be a vampire?

I'd have to check for tell-tale signs on my way round – if only I knew what they were.

There was also a very unpleasant whiff about the place that I couldn't quite put my finger on. Not that I really wanted to put my finger on anything in that strange house. The bouquet reminded me of dead flowers. Not one particular flower, but a whole bunch of them – a bouquet in fact. Get it?

I'm sure it was colder inside the house than it was outside because I could actually see my breath swirling around in front of me, on the still air of the room, every time I breathed out.

Curiously though, I couldn't see his!

A sudden chill shot down my spine and I could feel goose bumps starting to pop up all over my arms. I think I may have even physically shivered as if someone had 'walked over my grave' and it wouldn't have surprised me if my teeth hadn't started chattering as well.

So the curtains weren't there to keep the heat in, just the light out.

I could just about make out, by screwing up my eyes and squinting intently, that the floors were made out of dark oak. The large, heavy and imposing, furniture that sullenly occupied the room was also made out of dark oak and the bottom half of the walls were clad in, you guessed it, dark oak panels with the top half covered in dark flock wallpaper. 'Maybe Green' (similar to 'Could be Green' and 'Might be Green') but I wasn't sure – obviously.

Is he allergic to pastel colours?
Is he allergic to bright colours?

"Are you Mr Adams?" I enquired much slower this time and with a more baritone timbre to my voice. Not quite 'Butch', more 'Sundance' if you know what I mean. Although it was a bit late to ask the question now, I was already in.

"I a-m," came his measured reply – almost six inches.

His first name, however, was not Will.

"I believe you have a problem with a leaking roof?"

"D-o n-o-t b-e-l-i-e-v-e e-v-e-r-y-t-h-i-n-g y-o-u a-r-e t-o-l-d," was his very odd reply.

I wasn't quite sure how to respond without sounding completely stupid, so I asked him, "Are you telling me that this is all some kind of joke?" and so sounded completely stupid.

I wouldn't have minded, but it wasn't even April 1st.

"I n-e-v-e-r j-o-k-e."

I didn't have a problem believing that – believe me.

"So you do have a leak then?"

That's a leak and not a leek. Although to be fair he may have had both, but I was only interested in the former.

"Y-e-s."

"Can you show me where it is?"

"O-f c-o-u-r-s-e I c-a-n."

He didn't move.

Before I could, correctly, ask him, 'Please show me where it is,' which was after all, the phraseology I should have used in the first place, he turned around and started to walk away from me.

He must have had metal segs on the soles of his shoes because his slow, deliberate, and surprisingly short steps; c-l-o-m-p, c-l-o-m-p, c-l-o-m-p, were almost ear-splitting in the comparative silence of the room. There wasn't even a clock ticking.

There wasn't even a clock. Perhaps time didn't matter to the living dead.

With those short steps I now knew why it took him so long to answer the door.

Mr Adams's, not strange but weird, house was a three storey limestone building with a tiled roof. It was probably constructed around the turn of the 19th century by those well-known builders of the time; 'Sherlock Homes'.

I suspected most of the features were actually original, including Mr Adams by the look of him. The staircase certainly was, being an elongated helix that was also made out of dark oak – what else – and groaned with every footstep placed upon it. At least I thought it was the staircase.

We had to climb to the top of it, to the attic, where the alleged leak was said to be.

The hall and staircase were as dimly lit as the living room, but this time by grotesque, grotesque shaped wall lights with clear crystal pear drops that hung from them motionless in the increasingly oppressive stagnant air. I was beginning to understand exactly how a mole must feel like fumbling around in the dark. I laboriously followed Mr Adams up to the first floor. C-l-o-m-p groan, c-l-o-m-p groan went his footsteps, whereas mine just went groan, groan – a benefit of soft soled shoes.

On passing one of the bedrooms on this floor, I nearly jumped out of my skin.

A clever trick if you can pull it off.

I suddenly and unexpectedly came across another very tall, extremely thin man with a ghostly pallor and deeply sunken dark eyes that were accentuated by the dark circles that surrounded them, lurking menacingly in a doorway. Needless to say he was dressed from head to toe in black. There I said it anyway!

Is he a clone?

Are they the result of some secret scientific experiment that had gone horribly wrong?

The difference between him and Mr Adams was that his hair was almost down to his waist and so greasy that it hung pathetically in limp rats tails. He also had an alarmingly pained expression fixed to his face, not unlike that of a constipated baboon.

Don't ask.

I must have given an audible, if involuntary, gasp at the sight of him because the next words I heard were; "D-o n-o-t w-o-r-r-y," from that now strangely familiar and oddly comforting voice. "T-h-a-t i-s B-r-a-m. (As in Stoker! No surprise there). H-e i-s m-a-d."

Pot, kettle, black, was my immediate thought.

Then he slammed the door in Bram's face.

The house had now been upgraded from weird to plain spooky.

I was sorely tempted to ask why he thought Bram was mad, but I was intending to retire when I was sixty.

We proceeded painstakingly up to the second floor where I could hear the faint, but distinct, sound of classical music floating on the air, but I couldn't tell from witch (I know it should be which – just humour me) room. It was the funeral march to be exact – Chopin's of course, not Beethoven's. The sound was very tinny and extremely crackly so must have been coming from a 78rpm record being played on an old fashioned gramophone. I wondered if there was a Jack Russell Terrier staring into the trumpet shaped speaker as well.

Even in the increasing murkiness of the landing, Mr Adams must have seen the expression on my face because he declared triumphantly; "I-t i-s s-o r-e-l-a-x-i-n-g d-o y-o-u n-o-t t-h-i-n-k?"

Yeah, the ultimate chill out music – unless you're being cremated that is.

I just smiled. I don't know if he noticed it or if he even knew what it was, but he didn't respond to it.

The house was no longer plain spooky, it was now downright freaky.

We eventually reached the top floor and the attic. Here we discovered a female member of the family sat on the floor with her arms wrapped round her shins holding her bent knees close to her chest and her chin lodged in the valley between them. On our arrival she started to rock backwards and forwards manically whilst wailing uncontrollably.

A wailing banshee came to mind, but I don't know why.

"T-h-i-s i-s M-a-u…

I thought he was going to say Morticia.

"…r-e-e-n. S-h-e i-s a-n a-r-t-i-s-t."

Of course she is.

103

Morticia, sorry, Maureen was dressed from toe to head in black – naturally, her hair was black – unnaturally, and she wore thick black make-up – obviously. Needless to say she was also very tall and extremely thin with a ghostly pallor and deeply sunken dark eyes which, in this case, were accentuated by Goth make-up. I bet she still had dark circles surrounding them underneath it all though.

The room was littered with paintings all leaning haphazardly against the walls, and each other, lots and lots of paintings – and then some.

Some were large whilst others were not so large, but all of them were dark, very dark indeed. And that wasn't just their colour; it was their subject matter too. They were all, without exception, painted in shades of grey (fifty – approximately) with the odd splash of white for moonlit highlights and dramatic affect and only the signature was in a contrasting colour – crimson. Or was that the dried blood from the puncture marks on the neck of a female teenage virgin?

I'm sorry, I'm getting carried away. I certainly wished I could have been at the time – but not by any of them I hasten to add.

The themes of the paintings were demons, death, destruction and decay and probably some other words beginning with the letter d that I can't recall.

There were images of skeletal hands bursting out through freshly covered graves and strangling the poor innocent bystander; tormented ghouls floating aimlessly in the ether, whose faces made those in 'The Scream' and 'A Christmas Carol' look positively ecstatic in comparison; The Grim Reaper – with his crazed eyes staring out from underneath his cowl – brandishing a scythe that glistened in the eerie light of the moon.

It was a laugh a minute in that house I can tell you.

It was like waking up in someone else's nightmare.

Freud would have had a field day/week/month/year.

The house was now way beyond downright freaky and was now officially positively scary.

"S-h-e i-s c-h-a-n-n-e-l-l-i-n-g a-t t-h-e m-o-m-e-n-t."

Well it certainly wasn't BBC2 that was for sure.

He didn't need to point out where it had been raining in because I could see an ethereal shaft of light pouring in through the roof as if from some divine Super Trouper. There wasn't a plaster ceiling in the attic or any roofing felt either, just old roof tiles precariously attached to the exposed purlins. It was plain to see that there were at least three or four of the tiles missing at the apex of the roof resulting in the ingress of moisture – or 'raining in' to put it in the vernacular.

The paintings must have looked even scarier, if that was at all possible, bathed in real moonlight. I certainly wouldn't have wanted to be up there after dark. Not for all the tea in china, or even all the Guinness in Dublin for that matter.

'Run for your lives!' I wanted to scream. 'Save yourselves. You're all going to die if you're exposed to those 'unhealthy' vitamin D-infused rays of sunshine for too long.' But I doubted they'd have got the joke.

Joking aside though, it probably was the daylight that they were more concerned about – because they had a bucket to collect any rainwater.

"Have you contacted your agents and told them about the situation?" I enquired. A bit 'unconventional' he may have been, but I was sure he wasn't stupid.

"Y-e-s," was his short and not very helpful answer.

"And what did they say?"

"T-h-e-y w-i-l-l g-e-t s-o-m-e-o-n-e t-o f-i-x i-t."

"When did they say that?"

"T-w-o w-e-e-k-s a-g-o."

It hadn't rained during the last fortnight. (If you've been paying attention you'd have realised I'd already told you that).

But even so, that was totally unacceptable.

"That's totally unacceptable," (I was unanimous) I told him. "Who are your agents?"

He then produced, from his right hand trouser pocket, a scrunched up piece of lined paper, torn haphazardly from a notebook, on which he'd written, in black – obviously, spidery writing the details of the managing agents for the property.

"I'll contact them as soon as I get back to the office and explain that if they don't do the work within the next few days I'll have no alternative but to serve a notice on them."

I don't know if he was pleased by what I'd said or not because his expression didn't change.

Perhaps he'd been one of the first ever experiments with Botox.

When I came to leave the property Mr Adams went through the 'door ritual' again. The heavy velvet curtain was dragged away exposing the front door, the two heavy bolts clunked open as they were slammed back across the door and the key was hastily turned in the lock. I'd never seen him move so fast. Before I knew it I was back on the street – some things

105

never change – and before I knew it (for the second time), the door was well and truly shut behind me and I could hear the 'door ritual' being performed yet again.

I hadn't even had a chance to say goodbye.

It took quite a while for my eyes to blinkingly adjust to being outside again. I felt as though I was being blinded by the light. It was only then that I realised Mr Adams hadn't blinked the whole time I was with him. Not once.

Does he have any eye lids?
Are they stuck in place?
Is he in fact an alien?

So many pertinent questions, yet not a single relevant answer!

I contacted the managing agents, as promised, because 'a promise made is a debt unpaid', who informed me that they always had difficulty getting contractors to go to the house.

Why wasn't I in the least bit surprised?

"Why can't you get a contractor to go to the house?" I asked restraining a childish smirk. Even though I knew the answer, I still wanted to hear them say it. Rotten I know.

"Whenever anyone goes there, they get......"

There then followed a lot of contrived coughing, that wasn't in the least bit convincing. It was as though they were actually trying to avoid answering my question – surely not.

"Get what?" I pushed.

'Scared', was the answer I was expecting and, "Scared," was the answer that I got.

"What do you mean, sacred?" I was really enjoying this.

"Have you ever met the Adams family?"

"Yes," I said nonchalantly. "They may be a bit eccentric (a bit!), but otherwise they're harmless," I went on to explain.

"Try telling that to our contractors. Everyone who works for them has been to the property once and refuses categorically to go back again. They all say it's really...... Creepy."

"Pish," I proclaimed, "It's not creepy at all." Who was I kidding?

"They say it gives them the heebie jeebies and they're frightened that something will happen to them."

"Like what?" I encouraged.

"I don't know, they won't say."

I now realised that I was starting to get sucked into some hypothetical scenario that was not only ridiculous, but was just pure fantasy. The Adams's were innocuous enough for goodness sake and besides, if I'd never met them and they hadn't been as they were then I wouldn't have been able to tell you this bewitching (that's enough now) story.

So every cloud does have a silver lining.

"That's complete rubbish; I think that they all need to grow up," I admonished.

"But what are we supposed to do if no one will go there? It's an absolute nightmare." And those were her exact words.

"Unfortunately that's not really my problem," I said rather unsympathetically. "And it's certainly not an excuse for not doing the work either. I suggest if you can't persuade your existing contractors to go back then you need to get someone else to do the work."

"Are you going to go back?" was the acid response.

I hadn't actually thought of that. That'll teach me to be so smug in future.

"Of course I am," I replied trying to sound blasé even though I'd gulped quite audibly. "To check if the work has been done properly."

Or I suppose I could just ring the 'Troglodytes' and take their word for it. Except that I still didn't have a phone number. I didn't know if they even had a phone because I didn't see or hear one during my visit and even if they did, the number was probably ex-directory anyway.

A few days later, when I returned to the office from yet another life altering (the complainants, not mine – although I don't know though) visit, there was a message from the managing agents, left on my desk, to say that the work had been completed at the Adams's house.

Shock, horror (sorry – couldn't resist).

How they'd managed to persuade their contractors to go back I hadn't the slightest idea, but as long as the work was done properly, I couldn't have given a monkey's. That is, of course, unless they were different contractors altogether.

The very next day, yes I really was that efficient – once, I returned to the Adams's family residence to see if the work had been completed to my satisfaction.

It was drizzling intermittently as I set off for Greater Winlock and the combination of two weeks' dust and rain started to smear unsightly arched streaks across the windscreen of my car with each sweep of the wipers. Perhaps it was a, less than, subtle hint that my car needed washing.

By the time I got to Greater Winlock it was actually tippling down. So after parking the car I dashed (I've always been dashing) along the rain glossed pavements to the Adams's house, huddled under my inadequate umbrella for, not much, protection. By the time I got there it wasn't quite raining stair-rods, but it was heavy enough to bounce off the pavement and splash up around my ankles and soak my feet.

It's going to be easy to put the contractor's workmanship to the test today.

I rang the doorbell and in time honoured fashion I heard the 'door ritual' being performed once again.

By the devilish grin on his face I think Mr Adams was pleased to see me.

Okay, I'm making that bit up – he looked exactly the same as before, I just pretended that he was pleased to see me.

He did, however, have the same ghostly pallor, was wearing the same dowdy clothes and the atmosphere inside the property hadn't changed one iota.

Once inside he took my, now dripping wet, umbrella off of me, opened it up completely and placed it carefully on the floor. He obviously wasn't superstitious.

Like it's going to dry out in here, I thought.

He then took me, squelching all the way, up to the attic to show me where the work had been done.

On the way up I was surprised that there were no surprises this time. What I mean is; no one was lurking in any of the doorways; there was no muted music, of any genre, permeating the hall, stairs or landings; no one was wailing hysterically; and there was no sign of the paintings that had previously littered the attic. They had all totally vanished.

Perhaps they'd been requisitioned and were now part of a new exhibition at 'Tate Morbid'.

How odd.

Was I in the right house?

Maybe they'd realised that the contractors might be 'put off' if they encountered any of their aforementioned idiosyncrasies.

Wusses!

The work was completed properly and after that, I never heard from the Adams's again.

Da da dah!

Later that year, I could have sworn I saw at least two of the family featured in the Michael Jackson video *Thriller*!

Chapter Twelve

As part of the ridiculously complicated, scoring system for re-housing by the council – please don't ask me to explain it because I didn't really understand it myself, but keep that to yourself – the Environmental Health department used to inspect applicant's homes to assess their fitness. That's the home and not the applicant. Unfit properties invariably scored more points, and 'what do points make?' In this case it wasn't prizes, but they did propel the applicant closer to the top of the re-housing list.

My very first visit of this nature was to an address in the village of Dutton Pryors. The actual details I was given were: Miss M Bratby, The Cottage, Dutton Pryors.

How on earth was I supposed to find that? I was used to using an Ordinance Survey map, instead of an A – Z, by then, but that was going to be of no use to me at all in this case, because I didn't even have a road number.

"Don't worry," my colleagues tried, in vain, to reassure me. "Ask for directions when you get there. Someone is bound to know where it is."

"If you say so," I said completely unconvinced.

I admired their optimism, but I still didn't have their confidence.

When I arrived in the village the first thing I noticed was the small shop opposite the neatly mown, lush green, green.

Yes I could see it – thanks for asking.

The pond, next to the green, was replete with bulrushes, water lilies and ducks; two mallards, one pochard and a teal to be precise.

I had my 'I Spy book of Ducks' with me.

What an incredibly picturesque scene it was, the perfect picture postcard - 'weather here, wish you were nice' and all that. It conjured up thoughts of long hot summers, cream teas, lashings of ginger beer and the sound of leather on wood – that's cricket and not S&M you understand.

Someone in the shop is bound to know where 'The Cottage' is, I hoped.

I entered the shop to the sound of a tinkle – that being the bell above the door and not someone urinating in a corner. It was the archetypal 'Ye Olde Village Shoppe' with its low exposed beams, lumpy bumpy walls and creaking wooden floorboards. It was an absolute Aladdin's cave, a veritable cornucopia of delights and any other magical descriptor you choose to use. It must have sold practically everything the locals could have ever wanted, or needed, and some stuff that they probably didn't, but were persuaded to buy anyway. It had the staples – and other paper fasteners – of bread and milk as well as the practical like shoe polish and toothpaste, all crammed into every little nook and cranny.

That's cranny not granny – put your glasses on.

And there were plenty of them.

Again that's crannies not grannies, although there were probably plenty of the latter in the village as well.

It reminded me of the shop in 'Open All Hours'. I bet the prices weren't that dissimilar either – pro rata of course.

On hearing the bell a medium height and slightly overweight, white haired gentleman emerged from the back room dressed in a caramel brown overall.

"Good morning to you young sir and how are we today?" he asked jovially in a broad Shropshire accent.

I must admit, I was half-expecting a Yorkshire accent – with a stammer.

"Good morning," I said, "I'm very well, thank you," and I was. "How are you?"

"Mustn't grumble, mustn't grumble. But enough about me, what can I do for you on this fine day?" His enthusiasm was contagious.

"I was wondering if you could possibly help me. Do you happen to know where I would find 'The Cottage' please?" I enquired with more than a modicum of anticipation that my quest would soon be over.

"Don't know that I've heard of that," was his disappointing and unwanted reply. "Why do you need to know?"

"I have an appointment with a Miss Mary Bratby," I informed him.

"Oh the Bratby place, why didn't you say so in the first place? That'll be just over half a mile up the next lane on the left when you get to the crossroads, past the field on the right hand side with the Texel sheep in."

He could have just said sheep.

"You can't miss it; it's the thatched cottage with a white picket fence and a lilac tree by the gate."

111

Fortunately lilac was one of the few trees that I could recognise, so we were on to a winner with that reference.

"That's great, thank you very much, you've been most helpful."

"Not at all young sir, it was my pleasure. Now, is there anything else I can do for you while you're here? Anything I can get you? Perhaps you'd like some sweets for the car or a cold drink maybe? I'm never one to miss the opportunity for a sale," he said with a cheeky grin.

Arkwright was alive and well and living in Dutton Pryors!

How could I refuse such honesty?

"A packet of mints then please," I requested, as I would have felt guilty if I'd left without buying something.

"And while you're up there will you tell her I'll be round tomorrow morning with her groceries?" he asked as he supplied me with my change.

Yes, believe it or not, home deliveries are not a new phenomenon and they used to be achieved without the aid of a computer.

"Of course I will," I said accommodatingly and turned to leave.

"And thank you for supporting my business," were his last words called out to me as I left the shop.

I don't think he'd exactly made a fortune out of me, but I bet the knowledge that Miss Bratby was expecting a visitor was worth more than its weight in gold. I suspected that her business would be everybody's business in the village within the hour.

Following his very specific and dead accurate directions, I found 'The Cottage' quite easily.

That's one of the many things I discovered about country life; no one knows your address, but everyone knows where you live.

His description of the place was dead accurate too. It was indeed a thatched cottage with a white picket fence and a lilac tree by the gate, except that the lilac wasn't in fact lilac – it was white. Is that even allowed?

The whole scene looked as though it belonged on the cover of a chocolate box. Although I must admit, I've never bought a box of chocolates with a picture of a thatched cottage on it in my life. And besides, I'm usually far too busy trying to get inside the box to give two hoots about what the picture is on the outside anyway.

The one thing he did omit to tell me, however, was that the lane, beyond the – Texel – sheep, deteriorated into a narrow dirt track with a deep drainage ditch on one side that was lined with a thick bed of vicious looking stinging nettles. I was so grateful it hadn't been raining recently because I was sure that the track would have deteriorated even further; probably into a quagmire.

I had to be particularly careful parking the car, not that I'm not usually, so that; (a) I didn't fall arse over tit into the ditch and (b) there was enough room in case a tractor came thundering down the lane and wanted to get past – which would have been just my luck.

As I walked towards the gate I could see that the herbaceous borders, that surrounded the property, had been planted in a quintessentially English cottage garden style and an intoxicating scent filled the air. There was a profusion of different flowers in a kaleidoscope of colours that I didn't know the names of. That's the flowers and not the colours – unless of course they had Farrow and Ball names.

All the borders were edged with a selection of sea shells, some of which reminded me of the elegant fans used by the aristocracy in pre-revolutionary France.

'Who lives in a house like this?' – Miss Marple – that's who!

It looked idyllic and, for obvious reasons, it reminded me of the nursery rhyme:

Mary, Mary, quite contrary,
How does your garden grow?
With silver bells, and cockle shells,
And pretty maids all in a row.

And her name was Mary too – what a coincidence.

I tentatively opened the rickety old garden gate, that hadn't seen a lick of paint in many a long year – or a drop of oil either based on the noise it made – frightened that it might disintegrate in my hands.

Then it became all too clear why it hadn't been oiled. The sound of it squeaking open was an early warning signal for the geese that were congregated around the back of the cottage.

First I heard the distinctive honking noise they make as I turned round, after closing the gate behind me, followed by the unwelcomed sight of them waddling round the corner – at some considerable speed I might add – making a bee line for me.

You can keep your ferocious guard dogs as far as I'm concerned because geese are a much more efficient and effective deterrent any day of the week.

The sight of a dozen or more snapping beaks advancing in my general direction prompted me to leap back over the garden gate to the relative safety of the lane.

I say leap, it was more of a scrabble – even though I didn't get a double word score – in reality. But leap makes me sound so much more athletic.

Thank goodness the gate didn't collapse.

Even though I was now out of reach they still stood there defiantly rising up on their tip toes – well webbed feet – and flapping their wings furiously in a totally intimidating display of aggression. I bet Miss Bratby had never been broken into.

So what was I going to do now?

There was a gaggle of deranged geese between me and the front door, the door that I needed to be knocking on. If only I'd had my hover board with me!

Fortunately I didn't have to wait very long before my embarrassingly awkward predicament was resolved as Miss Bratby appeared at the door and wolf whistled to call them off. And miraculously they promptly obeyed.

Miss Bratby looked just like a farmer's wife. Well, my idea of a farmer's wife at any rate. She was quite stockily built – no I don't mean fat – and had a handsome face with a ruddy complexion. She was wearing a brown knee length herring bone tweed skirt, a hand knitted crew neck – very complicated pattern – fair isle pullover, brown brogues; a faded green quilted jacket with numerous threads of cotton dangling down where the stitching had come undone and a midnight blue – although it could have been black to be honest – silky headscarf.

As I said, just like a farmer's wife.

I'm stereotyping I know – because I'm actually typing with two fingers!

"Come along my dear, come along, they won't hurt you," she barked. I think she might have been a little hard of hearing.

That's easy for you to say, I thought. *It's not you they want to brutally mutilate then peck to death.*

I'm not exaggerating!

Cautiously I opened the gate once more.

"I heard the commotion so I knew that someone was about. Please close the gate behind you there's a dear," she requested as I walked apprehensively through it again. "I don't want the little darlings escaping you see."

*Little darlings, **little darlings**!* That's not the name that came to my mind a few minutes ago. A demon hoard, (I told you to put your glasses on), would have been a more appropriate description.

I turned around after closing the gate anticipating the 'little darlings' hurtling round the corner again, but no, they were obviously a well-trained and obedient 'demon hoard'.

"Have you come from the council?" she tentatively enquired.

"I have indeed."

"You'll have to speak up my dear I'm a little hard of hearing."

Not just a pretty face!

"I have indeed," I repeated, a little louder this time.

"I do hope you weren't frightened by my little darlings?"

"No, not in the least," I lied.

"Jolly good, they're so very useful you know. They provide me with eggs; they let me know when I have visitors and if I don't want any visitors, they chase them away."

I just smiled pathetically; I didn't know how to respond.

"Well then, come in, come in, don't just stand there looking like piffy on a rock bun," she said as she basically manhandled me into the cottage.

I had no idea what 'piffy' was, even though I apparently looked like it.

"So you've come to inspect my home have you?" she asked rather too cheerily.

"I have, yes. I believe you're hoping to be re-housed by the council?" I asked as she marshalled me through a convoluted obstacle course of 'clutter' into a large and cramped, chaotically arranged, living room. The distinctive waxy smell of furniture polish pervaded throughout.

Amongst all the 'clutter' was a drop leaf table with a set of four matching chairs in what could well have been walnut; an extending table with a collection of six mismatched chairs tucked neatly underneath it; two sideboards, one of which matched the extending table, and three display cabinets cram packed with 'stuff.' In the far corner was a beautiful roll top writing desk that was open exposing its drawers – I know, I shouldn't have been looking – and other secret compartments. I wondered if she still used it from time to time because there was a well-used ink blotter and fountain pen placed neatly on it.

The light from the two leaded windows, which looked out over the front garden, reflected off every polished surface in the room and illuminated the scene in a serene golden glow.

Scattered around the room, quite randomly, was a selection of clocks that included a grandfather and a grandmother clock, two carriage clocks, two anniversary clocks and a mantel clock. There was also a nest of three tables; a Welsh dresser; a dilapidated two-seater fabric sofa and one 'chesterfield' leather winged armchair. It looked more like an antiques showroom than a living room. And everywhere there were framed photographs covering every possible flat surface, but none on the walls strangely enough. Some of the frames were elaborately ornate whilst others were comparatively simple and plain.

Was it an indication of the importance of the individual in the photo? I wanted to know, but didn't actually ask.

Some of the photographs were sepia and quite badly creased, but most were black and white and in fairly good condition. It was like having a silent audience watching us.

There wasn't a television anywhere to be seen, but there was a big old radiogram which I suspected she would have listened to Jean Metcalfe presenting 'Two-Way Family Favourites' on the 'BBC Light Programme' on, back in the day. "No, not really, it's my family that want me to move," she replied. "I'd much prefer to stay in my precious home with all my valuables actually. Please sit down," she encouraged without pausing for breath.

I took a seat on the only part of the sofa that was minus any of her 'valuables' whilst admiring the multi-coloured hand crocheted antimacassars that bedecked it.

As I sat down there was an almighty twang.

"Don't worry," she said, "It's just one of the springs. It has seen better days."

Don't worry! The spring only nearly ruined my chances of fatherhood.

"So it's your family who want you to move?" I enquired as if she hadn't made it perfectly clear the first time.

"Yes, my nephews and nieces, they only want what's best for me."

How do they know what's best for you? I wondered.

"So do you want me to carry out the inspection or not, it's really up to you."

"I don't know. What do you think I should do?"

I could see she was in a bit of a quandary, but it was very difficult for me to give an opinion – because it really had nothing to do with me.

"Whatever you think is best," I said unconvincingly.

She thought about it for a moment or two then replied uncomfortably, "I really don't want to upset anyone, so I'll do what they want me to do."

It was her turn to sound unconvincing.

She didn't want to upset anyone – except herself that was.

I could tell that she knew it was a fait accompli anyway.

I'm sure that her family did have her best interests at heart, and God only knows she deserved a bit of luxury – if mod cons can be classified as a luxury – in her autumn years. I did, however, have a nagging doubt in the back of my mind that their actual interest was in her 'valuables', because they were in fact valuable.

Families eh!

Before I was allowed to inspect her 'precious' home Miss Bratby insisted that I had a cup of tea with her first.

"How do you take it?" she enquired as she got up from her chair.

"Milk, no sugar please," I replied as she went into the kitchen.

She entered the kitchen at 2.00 pm precisely.

How do I know that?

I know that because all the clocks in the living room started to chime simultaneously – and at the same time – in an alarming cacophony of sound.

I certainly wouldn't have appreciated that at 2.00 am that was for sure.

Fortunately there wasn't a cuckoo clock in the collection as well.

And thankfully it wasn't midday either.

In no time at all she returned carrying two large mugs of piping hot tea in one hand and a plate of yummy looking, and tasting as it so happened, homemade biscuits in the other.

The job did have some perks.

I tried to eat a crumbly, melt in your mouth, biscuit without making a mess and dropping too many crumbs, but I didn't succeed. Miss Bratby didn't seem particularly perturbed by my clumsiness and I certainly didn't need telling twice to help myself to a second one.

During our ensuing conversation she regaled me with tales of her life at the cottage, the cottage she'd been born in, 82 years previously – she honestly didn't look that old – and had lived in all her life.

There had never been a school in Dutton Pryors and so as a child she used to walk the two miles to the school in the next village each morning and two miles back home every afternoon; no matter what the weather. No mean feat believe me, because in the winter it could snow pretty heavily in that part of the world. No wonder she was still going strong. She was built of hardy stuff.

She told me that she'd never married even though she was engaged and proudly showed me the ring on her finger. Her face lit up with the pure pleasure of recollection. The ring looked like a brass curtain ring to me, but it had obviously served its purpose at the time and I could tell meant the world to her.

Her childhood sweetheart had gone off to war in 1917 and never came back. It was obvious that she still loved him dearly and was waiting to be reunited with him, wherever and whenever that would be. She then leaned over to the sideboard behind her and passed me a faded black and white photograph, in an ornate silver frame, of 'Bert'. When I handed it back to her, after agreeing how handsome he was, she stroked it affectionately, but unconsciously, throughout the rest of our conversation.

She also told me that over the years, as all her younger brothers and sisters had died, she'd inherited their 'valuables' and so all their furniture, memorabilia and other knickknacks were now stored in her place.

That would explain why it was so cluttered and cramped.

During the course of our 'tea break', she brought out a selection of photograph albums and scrap books that were filled with pictures of people long dead in life, but not in her memory or heart. They traced her family tree back to her paternal great grandparents. Underneath each picture, written in the most beautiful copperplate hand, was the name(s) of the person(s) in it and the date it was taken. It was a complete history lesson about the Bratby family and extremely fascinating it was too.

These were the days when we actually had time to talk to, and care about, people as individuals and human beings. They were not just some, meaningless, arbitrary target that had to be achieved. I'm talking about the quality of a visit and not just the quantity of them.

'I am not a number,' as The Prisoner used to say.

When I eventually got round to inspecting the property I was amazed how she could have lived so frugally for so long.

She didn't have an inside toilet and the outside convenience she did have was not that convenient. It was actually at the bottom (how appropriate) of the back garden, just beyond the vegetable patch that was brimming with a large variety of produce that could have fed a small African nation for a couple of months, and was 'bog' standard by anybody's standards.

See what I did there.

But at least it was for her sole use as it was guarded by the geese.

I really wished there'd been more than just her there so she could have said it was for 'our sole use.'

All very handy for the middle of the night – not! She must have had a Guzunder – a chamber pot to you – in the bedroom, in case she got caught short in the middle of the night surely?

Her kitchen was basic yet functional and consisted of a very old, but still in good working order, range, a slightly chipped, but still white, Belfast sink with only a cold water tap over it – the kettle on the range provided her with hot water – with a wooden draining board and a couple of floor cupboards to store her pots and pans in as well as her food supplies. The tops of them were used as food preparation surfaces. She didn't have a fridge, but did have an old fashioned meat safe instead.

Everything was spotlessly clean, she was obviously very house-proud. There wasn't a bathroom and I'm quite sure that in the 'old days' the whole family would have used a tin bath, one after the other, in front of the fire. Bagsy not going last!

There wasn't any central heating either, only an open fireplace in the living room, and of course the range in the kitchen for heat.

The staircase that led up to the first floor was steep and rickety and the threadbare carpet, that just about covered it, was rucked up in several places creating what I considered to be a serious tripping hazard. Yet she'd never fallen down the stairs, not once – ever.

There were three bedrooms upstairs. The second was entered by walking through the first and the third was entered by walking through the first and second.

Not particularly convivial for privacy I admit, but it was a fairly common design feature of properties of that age in that area.

The parents would have slept in the middle bedroom with the boys in the first and the girls in the third. This was to prevent any incestuous activities between siblings, during the night, which were more common than you might imagine at the time the property was built.

The upstairs consisted of another maze of more furniture and 'stuff' to get to the third bedroom where Miss Bratby slept. I guess 'Old Habits Die Hard' (film number 19 in the franchise). Her bed was covered in a primrose candlewick bedspread that had a delicately embroidered pyjama case strategically placed in the middle of it. I suspected that she still slept in that bedroom so she could see all of her 'valuables', stored in the other two rooms, at least twice a day.

Most of this furniture must have been dismantled and then reassembled – by a child of seven I would have imagined – to get it up there, because there was no way it could have been taken up the stairs in one piece. There were three wardrobes, each made out of a different type of wood, with a different one in each room. One of them was considerably larger than the other two and I was convinced that if I'd waded into it I would have eventually emerged in Narnia. There were also two full length mirrors, one of which was quite masculine in appearance being oblong and rather plain whilst the other was very feminine having an oval shape to it and a filigree design carved around the outside of it. I was equally convinced that if I'd stepped into it I would have met Alice, The Red Queen, The Unicorn and all of her other 'Wonderland' chums. The other furniture included three ottomans, four tall boys (and one short girl – just being silly) and a conglomeration of arm chairs, occasional tables and other 'odds and sods'. And more photographs. Again, I wondered if the upstairs portraits were not as important as those displayed downstairs.

Surprisingly there weren't any clocks upstairs. Either she could hear those downstairs or she had a finely tuned natural body clock and was at one with nature. I'd wager that she could tell the time by looking at the position of the sun in the sky on a sunny day – one of those weird and wonderful country ways.

I suppose you'd expect the rooms to smell musty with all that old 'stuff' in them, but they didn't. In fact the distinctive waxy smell of furniture polish permeated the upstairs air, just as it did downstairs.

When I'd finished my inspection I sat down with her and explained, as gently as I possibly could, that by modern standards the cottage was in fact unfit for human habitation.

"Oh, I do realise that dear," she acknowledged, "But I do so love it here. All my memories are here."

For the first time since I'd met her she looked vulnerable, and then she started to cry. Quietly at first but then it developed into wracking sobs. I tried to console her as best I could, but she obviously needed to let it all out and eventually she did calm down.

"When I go into sheltered accommodation I've been told that I'm only allowed to take two small pieces of furniture with me and a few other bits and bobs," she said still sniffling. "How am I supposed to choose what to take with me from all of this?" she gestured with a large sweeping motion of her arm.

I didn't have an answer.

"What about my beautiful garden? It's my pride and joy and I love pottering about in it and growing things. What will I have to do there?"

I still didn't have an answer.

"Also I'll have to leave Henrietta, Harriet, Hermione, Deidre, Daphne and Doris behind too."

At first I didn't have the slightest idea who she was talking about.

Were they imaginary friends?

Perhaps they were dead relatives whose ashes were buried in the garden.

Then it suddenly dawned on me. She was referring to the geese of course.

Only half a dozen?!

She must be mistaken.

I was convinced there were more than six when I arrived. I reckon there was a whole squadron of them.

"Surely someone will look after them for you?" I enquired optimistically.

"Oh yes, young Mr Young said that he'd have them. He owns the farmhouse on the other side of the village."

It turned out that YOUNG Mr Young was a mere 75 years old!

"And I'm sure your family will bring you back to see them from time to time," I suggested.

"Oh, I'm sure they will," she said resignedly.

I eventually left her, still clinging onto the photo of Bert, and feeling a bit more positive about her predicament whilst I felt totally perplexed by the whole situation and as a result forgot to tell her about her groceries.

Miss Bratby was eventually re-housed in a modern, purpose built, state of the art sheltered accommodation complex. She had, by all accounts, a beautiful little flat with an inside toilet, a modern bathroom and kitchen with hot running water and central heating throughout. There was also a communal lounge where residents could meet up for a chat or a game of cards and evening entertainment was provided at least once a week.

Perfect you would think.

Unfortunately Miss Bratby went down with flu within a fortnight of moving in that rapidly developed into pneumonia and then double pneumonia. Within a month of her leaving her 'precious' home, and her 'valuables', to please her family, she was dead.

Although she was built of hardy stuff, she was obviously not used to the so called 'luxuries' of life, and so the bacteria, that caused the pneumonia, and had probably been passed on from one of the other residents, quickly multiplied in the warmth of her new surroundings.

I'm convinced that she'd have fought off the infection if she hadn't moved there. Then again, she wouldn't have picked up the infection in the first place if she hadn't moved there.

I doubt if she'd ever had flu in her life before.

I felt really guilty for a long time after she died that I'd in some way contributed to her demise and writing this now, and remembering her, has made me realise that I still do a bit actually.

Should I have advised her to ignore her family and do what she wanted to do?

Was her time up anyway?

I know I'll never know the answers.

You think you're doing your job and helping people, but it's not always the help they really want or actually need that you're offering.

Her family thought that they were doing what was best for her.

Makes you wonder who really knew what was best in the end.

I only hope that she was reunited with her 'Bert' at long last.

Chapter Thirteen

After I'd finished dolloping the dye into Mr Head's (you remember him from Chapter One?) cesspit – and stopped daydreaming – I made my way next door to his neighbour's house to explain the whole damn sorry affair to him.

Mr Riley and Mr Head were as different as chalk and cheese.

One was quite palatable, whereas the other most definitely wasn't. Not by a long chalk – cheesy I know!

Mr Riley was tall and broad. He had a crew cut hairstyle, a weathered face with a chiselled jawline and piercing electric blue eyes. He was what writers of romantic fiction would call, ruggedly handsome. I on the other hand would call him a bruiser of a big butch bugger. His handshake was warm and friendly even if it was bone-crushingly firm. His crew – to match his haircut – neck khaki t-shirt that matched his cargo pants, clung effortlessly to his muscular torso and left very little material to spare. There wasn't a gram of fat on the man. It did, however, manage to allow the tattoos on both his biceps to peek out from underneath its short sleeves, although I couldn't make out what they were of. He was obviously the rugged outdoor type, a bit like me really – well in my dreams anyway. He could quite easily have passed for an all American G.I. – even if he didn't have the accent.

Mr Riley was very understanding, very apologetic – even though I hadn't established whether it was his cesspit leaking or not – and very accommodating with his help in the matter.

He hadn't realised there was a problem and actually took me to his cesspit and opened it up for me so that I could put some blue dye in it.

So it was yellow dye in Mr Head's cesspit and blue in his Mr Riley's. Two days later I returned to see what had happened.

Yellow dye was bubbling up steadily out of the ground and wending its weary way gently down the lane.

Yes! I clenched my right fist and pulled my arm down in a quick sweeping motion.

It was Mr Head's cesspit that was leaking and it was him who was in Shit Street – or Lane if you're going to be pedantic.

I know it was pretty petty of me, but I can't abide people who are so sanctimonious and think they know it all, when in fact they don't.

I couldn't get to his front door fast enough. Unfortunately though there weren't any cars parked outside the house, so unsurprisingly I got no answer when I rang the doorbell – which was surprisingly just a simple chime and nothing at all pretentious. I was desperate to rub salt into the wound, or little prick if you prefer.

I'm certain that you would have wanted to too (not a ballet skirt). I get the sense that you don't like him either.

I found out from Mr Riley that it was in fact Mr Head's second home – I should have guessed – and that his main home was in London.

So he'd probably scarpered back to the smoke when he saw the yellow dye trickling down the lane.

Even though I didn't have his London address I did serve a notice, requiring him to repair the cesspit, on him at his second home, and after numerous threats of court action he did eventually comply with it.

Bet he would have expected immediate action if it really had been his neighbour's fault.

Bloody hypocrite! (What have I told you about your language?).

Believe it or not, earlier on in the same year I'd been called out – by a distraught young lady – to yet another drainage problem. And no her name wasn't Ripley, it was Brewer.

This time it was to a property where the sewage was going down the toilet much slower than it usually did, and the manhole in the back garden was starting to fill up with – well I'm sure you can guess what with.

The estate that Mrs Brewer lived on was only five years old so I managed to get the plans of the drainage system from the building control department. I would need to establish if there was more than one property affected by the blockage in case I had to serve any notices.

It was a stiflingly hot summer's day, the first of its kind that year, and as a result I was wearing sandals (no socks please, I'm British). Stupid I know – on more than one level.

I could see Mrs Brewer walking up the hall towards me through her, full length, frosted glass, front door, before she actually opened it. She was probably in her mid to late twenties, average height, average weight, but above average looks – way above. She wore a simple navy blue cotton dress with wide straps and a sweetheart bodice that had a liberal smattering of what looked like, and certainly smelled like, banana on it.

Tres de rigueur.

Either she was a very messy eater, or there was an infant involved. She wore nothing on her red toenail painted feet and brushed a stray wisp of her shoulder length blond hair back behind her ear as she greeted me. "Good morning," she said with a distinct look of relief emerging on her make-up free face. "Thank you so much for coming so quickly." In the background I could hear the contented gurgling sounds of a young child.

I would write down what those sounds were, but I'm not actually sure of the correct spelling. It was something like; goo, gar, blur, blub, blub – or 'words' to that effect. There may have even been a thrrrpp in there as well – unless of course that was something completely different.

I followed Mrs Brewer back down the hallway, being careful not to step on the toys that were strewn liberally over the beech effect laminate flooring, whilst admiring the pictures on the wall. One of them was obviously a professional studio posed photograph of the family; mum, dad and baby Rachel. The other was a picture of prints, not the artist formally known as symbol – or Tafkap to be precise – but of Rachel's feet and hands.

When we entered the kitchen, sat in her high chair was the cutest little girl imaginable. Okay – so all infants are cute. She had a Winnie the Pooh bib around her neck, a matching Winnie the Pooh beaker in one hand and the remains of a squished banana in the other. She was smiling broadly, which created dimples in her little cheeks and exposed her one incisor tooth, whilst she munched contentedly on what was left of the banana. Most of it looked as though it was caked in her thick blond hair, that which wasn't on her mother's dress anyway. She seemed happy enough though and her mother didn't seem too bothered by the mess either.

"I'm so worried," she explained. "Rachel has just started walking and I would dearly like to let her out in the garden during this nice weather, but I can't."

It was then that Rachel lobbed a golf ball sized mushed up banana missile in my general direction which I managed to, just about, dodge. Her aim was pretty good it must be said. I could see that there was a future international netball player in my midst.

"I'm terribly sorry," declared Mrs Brewer, obviously embarrassed as the colour rose in her cheeks, as she gently reprimanded Rachel. She looked mortified – Mrs Brewer that is, Rachel just thought it was funny and laughed cheekily.

"Not to worry, no harm done," I tried to reassure her. "It could have been worse."

Little did I know that things were about to get worse – a whole lot worse. Worse in fact than the worst thing you could ever imagine made even worse by something twice as worse happening at the same time.

I left Mrs Brewer and Rachel to their 'elevenses' and went into the back garden to take a look down the manhole. All looked perfectly fine until I got closer to it. As I trod on the ground next to the manhole, the aluminium cover lifted off and raw sewage disgorged itself and before I knew it, I was up to my ankles in it.

Please, do not try this at home.

I bet none of those poos had names, Winnie or otherwise.

I now knew how Mr Grimshaw must have felt walking round his flat and it wasn't a pleasant feeling I can tell you.

Mrs Brewer must have seen what had happened because within minutes she was out with a bowl of 'baby safe' hot water (temperature checked with her elbow I suspected) laced with disinfectant, for me to wash my feet and sandals in.

It was now my turn to blush. I could feel the colour rising high and hot on my cheeks – all four of them.

I was sure that she was being generally considerate and had my personal welfare at heart and it wasn't just because she didn't want me treading (sh)it through the house. Although I was fairly sure she didn't.

Five minutes and a clean pair of feet and sandals later I was looking at the plans which indicated that the next manhole in the system was in the garden of number 6. Mrs Brewer lived at number 4.

When I went next door, reeking of what was probably Dettol, to ask if I could take a look down their manhole, I was politely informed, "There isn't one."

'You cannot be serious,' my John McEnroe side wanted to screech. But instead I went with the much calmer, "But there has to be, it says so on the plan."

"You are more than welcome to come and take a look," said the neighbour, obviously recognising my distress, "If you think it will help."

"It's not that I don't believe you," I said – hopefully convincingly, "It's just that I don't understand."

"I realise that."

I was hoping that that was the believe bit as well as the understand bit.

"Please come on through and see for yourself."

There was no manhole in the garden.

Of course there wasn't.

I knew she wasn't lying. I'd just hoped that she'd somehow missed it because there was an enormous shrub hiding it.

Now what? I was totally flummoxed.

I spent many a long hour that day knocking on doors, explaining the situation of the blocked drain and not knowing where everything drained to. Some of the properties had manholes that obviously didn't belong to the system I was investigating, but eventually I did find the ones that did.

The drainage plan bore no resemblance at all to the actual drainage layout. It was about as useful as a paper chastity belt in a fire.

In the end I established that there were three houses affected by the blockage and subsequently served notice on the three owners. The notice gave them 48 hours to clear the blockage or the council would do the work in default and recharge them. Inevitably owners would usually leave it to the council to do the work because it was easier than finding and agreeing on a contractor to do the work – and this case was no exception.

When the system was finally unblocked we discovered that it was a disposable nappy that had caused the problem. I discreetly let Mrs Brewer know because the other properties didn't have any babies in residence.

Mrs Brewer assured me that she never put nappies down the toilet, so we could only surmise that one of her neighbours had had visitors who had. (Three hads in one sentence – is that even legal?). Or perhaps Rachel had managed to do it!

Chapter Fourteen

Man has a veracious appetite for fuel. Not for him/her personally, but for all the gadgets and gizmos that are now part of everyday life, and we couldn't possibly live without, that need power – in the industrialised world at least.

Although you could say that man has a veracious appetite for personal fuel as well, in the form of food, that has led to the obesity crisis that we are now faced with - again, in the industrialised world at least.

In 'The Good Old Days', a veritable cornucopia of flowing electrical charge in the UK came from coal fired power stations and new supplies of the fossil fuel – to 'keep the home fires burning' – was always being eagerly sought after.

There was a small plot of land, about the size of six football pitches stuck together (2 x 3, not 6 in a row) – give or take a rugby pitch – on the outskirts of Idsall that had been identified as having coal beneath the ground. The site wasn't large enough to warrant a traditional full scale coal mine so the industry wanted to establish an open cast mine there instead – similar to a quarry.

There was, however, one ginormous problem with that proposal. Idsall was a quiet little village. In fact it was so quiet that at night you could hear an ant sneeze – but strangely enough, never an uncle – and any 'industrial' noise would have caused a considerable nuisance to the residents. So it fell to the council – and in particular the Environmental Health department – to establish a case for objecting to the planning application. This would involve demonstrating the potential noise nuisance that would be created by the mine and the subsequent repercussions. Not an easy task.

Northford Council was only a small council both in terms of its population and its staffing levels – even if it did cover a large geographical area. As a result, it only received a small income from the government's Standard Spending Assessment (SSA) and from local taxes.

It, therefore, only had a small budget for the purchase of any specialist equipment especially the specialist noise monitoring equipment that was necessary for such a big and complex job. Fortunately though, a nearby authority – that had the necessary equipment – kindly said that they would let us borrow theirs, free of charge.

What a stroke of luck.

The plan was to carry out twenty four hours of noise monitoring at five different locations around the proposed site over five consecutive days, Monday to Friday. Realistically there would only to be twenty three hours of monitoring at each location as it would take an hour to shut down the equipment and then move everything from one location to the next before setting it all up again.

We hired a small caravan to store the expensive, delicate and bulky equipment in. (Modern noise metres are so much smaller and more robust, but are still very expensive). The caravan would also serve as shelter for the officer who was on duty with the equipment as well. A suitable vehicle to tow it all around was also hired.

I wouldn't have described the caravan as small, because that wouldn't be accurate. It was tiny. So tiny in fact that it would have even been a squeeze for Ken and Barbie.

Although in all fairness, it didn't really need to be any bigger for our purpose I suppose. However, I wouldn't have fancied spending a week's holiday in it at a wet and windy, run down, seaside Caravan Park with a premenstrual wife and a bored teenager – in October. Not that that particular delight was on offer anyway. For starters I think it would have blown over, especially if the big bad wolf had huffed and puffed, and the sound of rain on the 'tin can' – which essentially is what it is after all – would have sounded like darts being thrown at it and driven me up the wall.

The caravan had everything you could ever need though – just smaller. It had an itsy bitsy teeny weeny w.c. compartment – with just enough room to 'drop 'em' – with a chemical loo; a four ring, single oven cooker in the kitchenette area, which also had a table and two bench seats that folded down into a single bed. And in the living area there was another table and bench seats that folded down into a double bed. It was all very cosy – with just one person in it. It was quite simply cramped with two, but with three? What's the next step up from cramped?

We manned – womened/personed, if you want to be politically correct – it in shifts throughout the twenty three hours. During the day one officer would be alone in the morning and another in the afternoon. However, at night, two officers would be there together – for something called 'Health and Safety'. I doubted it would ever catch on!

You can imagine the politics of who wanted to work with whom on the night shift. Or should I say who didn't want to work with whom. I just prayed that the person I was on with hadn't had a curry for dinner that evening, because there would be no escape from the consequences.

The night shift finished at 07:00.

Don't you think it's funny when people say in the morning as well when they've already said a.m. or are using a twenty four hour clock?

No?

I do.

At that time everything had to be stopped, packed up, moved to the next site, unpacked, set up, and then started up again by 08:00 a.m. in the morning.

An hour sounds like plenty of time, but it was always a hectic race against time, which luckily we won, every time – but only just. Three 'times' in a sentence this time.

The morning after the night the boss had been on duty he couldn't get away from the place fast enough. When my colleague, Jill Peckham, and I arrived to move the equipment he was out of the caravan like there was no tomorrow. As he left he shouted out his goodbyes, which echoed several times around the sleepy hollow; slammed the car door; revved his car engine like an F1 driver on the starting grid and then left with a squeal of tyres in a plume of white exhaust smoke.

All the time the monitoring equipment was still running, which caused a not inconsiderable spike in the noise readings that could have affected the final result. Fortunately, because it was only one abnormality – the reading not the boss – of less than a minute, in a total of 115 hours of monitoring, it wasn't too disastrous.

He didn't quite get it, poor love.

Jill was our resident 'noise expert' and a formidable force to contend with as well. Her personal mantra was; 'faults I may have but being wrong isn't one of them'. You never got into an argument with her unless you were absolutely sure of your facts, and if you were then she wouldn't get into an argument with you.

On the last full day of monitoring, when I was on the morning shift, I was starting with what I thought was just a cold. After we'd set everything up Jill left me to it and I sat down on the double bench seat and started to shiver.

Then I was boiling hot.

I was starting with the flu.

No not man flu – real flu.

Basically I just slept the whole time I was there. If anyone had come to nick the equipment I wouldn't have known – or cared to be honest. If the equipment had failed in some way, I wouldn't have known that either and that was the whole point of being there in the first place.

This was in complete contrast to Ron Wright, the Principal EHO, who was so petrified of it all going completely tits up that he just stared at the equipment the whole time he was on duty.

Ron was in his late 50s when I started at Borntown and was the perfect English gentleman He always wore a suit, had thin wispy white hair and an impeccable 'BBC' speaking voice – and would do anything for you. Well perhaps not anything, but it certainly seemed that way, because I never heard him refuse 'a favour'. One of his foibles though – I won't list them all – was his 'addiction' to Benylin. He would even take it when he hadn't got a cold as the number of empty bottles strewn across the back seat of his car was testament to.

When Jill came back to take over at lunchtime she took one look at me and declared, "You look bloody awful."

I put that down to the flu, but didn't ask for confirmation – just in case it wasn't.

After that I went home, crawled into bed and didn't get out again until the Monday morning – other than for sustenance and going to the toilet of course.

Although the whole process was a bit manic and a steep learning curve for everyone in the department, it was an incredible experience for me and one that has never been repeated since. Not only that, but the overtime was brilliant too and I bought my very first microwave oven with the proceeds.

'Is that all?' I hear you scoff.

Yes, because they'd only just come on the market and were much more expensive than they are nowadays.

When we finally presented our case for refusing planning permission to the committee they wholeheartedly agreed with and accepted our recommendations without question and flatly refused it.

So it was yet another job well done.

Chapter Fifteen

When you're dead and buried you'd think that that was it, you're dead and buried. Well that's not necessarily the case, because under certain circumstances bodies can be exhumed.

Most people don't get to witness an exhumation in their lifetime and hopefully most people aren't the subject of one after it either. But then I guess most people wouldn't actually want to attend one either. You can't exactly buy tickets from Ticketmaster (other agents are available) for them.

Not many EHOs get to observe an exhumation during their careers and having been one of those who did, I suppose you could say that I was one of the lucky ones – from a professional point of view at least. Yes, I've actually seen one in the flesh – as it were. Even though, thankfully, there was no flesh to be seen.

'Exhumation' – in case you were interested – is the removal of human remains, either in the form of a body or cremated remains, from their place of interment. So you might have seen a dramatisation of one on programmes such as 'Waking the Dead' or 'Silent Witness'.

On the telly exhumations are carried out because the investigating detective wants another post mortem carried out on the body or a DNA sample taken. This is because, in the last twenty minutes of the programme, there has been a significant twist in the plot and new, previously undiscovered, clues have come to light. As a result our intrepid hero believes that they will now be able to nail the murderer with the vital piece of new evidence even though the crime was committed many years ago.

A bit of a spoiler I realise, but there you have it.

In this particular case it wasn't anywhere near as dramatic. The relatives of the deceased were moving to Wales and wanted to take the body with them.

In truth, exhumations are pretty rare and are only usually allowed in exceptional circumstances, but can occur for a number of reasons including to:

- move the body from the original grave to the site of a new grave (as in this case)
- repatriate the body overseas to be buried along with other family members
- deepen an existing grave for a further burial
- send the body for cremation
- be examined forensically as the result of a court order (as in waking the dead)

But it isn't as simple as just digging a body up. To carry out an exhumation you have to go through the rigmarole of getting a licence, and although the licence is free, the financial and psychological costs of the actual exhumation can be extremely expensive.

An exhumation must take place within 12 months of the granting of the licence and the time, date and place of the exhumation must be notified to the local EHO at least five working days prior to the event. Arrangements regarding the transport and storage of the remains between the period of exhumation and reinterment should also be notified to the EHO.

So what was my role in the one I attended?

When the request for the exhumation was received I had to visit and inspect the grave of the deceased (well it would hardly have been the living now would it?) in question and obtain any further information I needed from the management of the cemetery and the undertakers involved.

The exhumation could not take place unless I was present to supervise it. (Actually any EHO would have been fine; it didn't have to be me). This was to ensure that all procedures were complied with and everyone present showed suitable respect for the deceased and most importantly that public health was protected.

If the remains were going to be reinterred in the Borntown area, which they weren't, then I would also have had to supervise the reinterment as well. But because they were to be reinterred in another local authority area, then I ensured that the receiving local authority had all the necessary information for the reinterment. If, at any point, I decided that the removal of the coffin was in any way a health hazard then I could stop the exhumation immediately – the power!

During the course of my exhumation, or more correctly the exhumation I attended, I ensured that:

- only the necessary people were present
- the correct grave was opened
- the exhumation commenced as early as possible in the morning to ensure maximum privacy
- screens were placed around the existing grave to protect the exhumation from public view and to guarantee privacy
- all the workers were wearing disposable protective clothing, including gloves, overalls and face masks, (but not funny ones obviously), and they were all disposed of safely following the exhumation, (the protective clothing, not the workers)
- everyone present showed due respect to the deceased, (hence no funny masks), and the adjoining graves
- the name plate on the casket corresponded to the name on the licence
- all the remains and pieces of the original casket were placed in the new casket which was made of timber, tarred on the inside, followed by a zinc liner and then covered inside by a leak-proof membrane. The zinc liner was sealed with isopon. The new casket was eventually screwed down and a new nameplate stating the name and date of death of the deceased was attached
- a supply of disinfectant was available (it used to be lime)

I told you it wasn't simple.

By the way, I didn't make that a list of bullet points because I was running out of time. I just thought it would be easier to read.

The exhumation I attended was held at the crack of dawn (don't be rude) just as the sun was rising. Except that on that particular morning the sun didn't exactly rise, instead it sort of dragged itself, probably screaming, lethargically over the horizon and all the time I attended the exhumation it managed to successfully keep itself hidden from view behind a blanket of sinister looking cumulonimbus clouds as though it didn't actually want to witness the macabre scene below.

I knew exactly how it felt. I desperately wished that I was concealed by a blanket as well – the one on my bed.

In other words it was a typically miserable autumnal morning in November. Not only was it dark, but it was cold, it was wet and there was a stiff breeze.

What other kind would you expect under the circumstances?

It wasn't raining heavily, but the wind that whipped through the skeletal trees made it seem worse than it actually was. It was that horrible fine drizzle that soaks you through almost instantly and leaves you feeling cold, damp and miserable.

Why oh why couldn't it be held on a warm summer's morning in July?

The deceased had been buried for just over six years and under normal soil conditions their coffin would have probably rotted and collapsed and as a result there would have been the potential to see a decomposed body, or maybe even just a skeleton. Not a pretty sight at any time of the day and certainly not in the murkiness of that particular morning. All it needed was dry ice billowing over the ground, the sound of a wolf howling in the distance and a flash of lightening illuminating the sky and I would have been appearing in my very own Hammer horror movie.

The grave had been dug out the evening before we arrived and the coffin was left with a thin layer of soil covering it. Well mud actually, because the ground in that area was very clayey. This was not good news for the grave diggers, but very good news for us because clay, to a certain extent, preserves and so the coffin had not rotted, even though it had cracked, but didn't disintegrate when it was lifted out of the grave and so the body was not exposed. The smell, however, was awful. To coin that well know turn of phrase for something that smells horrible; it smelled as though something had died. I know it's a cliché, but it was an extremely apt one in that case.

The new coffin, the undertakers had brought with them, was one size larger than the original and so the old one fitted perfectly inside the new one. Then when the deceased was finally taken away I started to slosh disinfectant liberally all over the site and my work there was done.

Chapter Sixteen

Food complaints were traditionally – and I suppose still are to a greater or lesser extent – the staple diet, the bread and butter (all puns intended) of environmental health work.

One of the first food complaints I ever dealt with was concerning the discovery of a pubic hair in a burger. Well that was what the complainant claimed it was anyway, and I mean that both in terms of the hair and the pulverised pink pate (well pulp really) they presented me with. It certainly didn't look like any burger I'd ever seen before or one that I would ever want to eat for that matter.

Seriously though, I couldn't understand why they were so adamant that it was a pubic hair – unless of course it was one of their own.

I know, it doesn't bear thinking about, especially if they'd been bare with the burger – if you know what I mean. Luckily for you, this isn't a pop-up book. And it wouldn't be the first time, or the last, that someone had 'manufactured' a food complaint. (See what I did there?).

I reluctantly took the food (and I use the word loosely) off of the complainant and said that I would contact the manufacturers to see if they could get to the bottom of it – even though it wasn't rump steak.

The offending, and it really was offensive, item was from a packet of frozen burgers and not from a take away restaurant – just in case you were wondering.

A few weeks later I heard back from the manufacturers who informed me that on the face of it (literally) the hair was a bovine (beef) eye lash which would have got into the meat when the face of the animal was being boned out.

You mean you hadn't realised it was that type of meat that went into a burger? Sorry if I've shattered any illusions you may have had of what some burgers actually consist of.

Then there was the blue waterproof plaster that was found in a packet of crisps.

I always thought that the little blue sachet contained salt to sprinkle on the aforementioned thinly sliced fried potato snack.

Not in this case. The blue package contained puss, and a considerable amount of it.

Now the only time that a puss infested blue waterproof dressing is acceptable in food is when it's in custard – because no one would realise once the plaster had been fished out.

Joke! Honest!

But one of the most unusual complaints I ever received was concerning a clear plastic tooth pick that was found in a bottle of whiskey. Not the greatest idea for a free gift in the history of the world it has to be said. I think I would have sacked the advertising agency in question for that particular faux pas.

Unfortunately for the complainant, the tooth pick was discovered when it was lodged in the back of her throat.

When Mrs Wattleworth met me at her front door she seemed more than a little surprised by my presence even though she'd requested it. She was probably in her mid to late 70s, although, to be honest, I'm not very good at ageing people. Her eyes were the washed out grey/blue of a wintry sky, her snow white hair (not Snow White's hair) was neatly piled up on the top of her head in a bun and she smelt of mothballs. Well reeked of them actually.

I hadn't realised that they were still available. Perhaps she'd been stockpiling them and there was now a national shortage.

She lived in a ground floor flat with her pet Chihuahua – Elvis, who she was carrying lovingly in her arms.

Honestly, without a word of a lie, it was a ground floor flat.

Elvis was certainly no *Hound Dog*. I'd seen bigger cats, and I don't mean lions or tigers either. Or panthers or cheaters etc., etc.

How did I know the dog's name so early on in the encounter? From the 'bling' diamante letters stuck on its patent leather (plastic really) collar. I say 'it' because you would think that Elvis was male, but no, Elvis was a bitch and I mean that in the 'female dog' sense of the word and not a 'malicious, unpleasant, selfish person'.

Having established who I was Mrs Wattleworth hobbled off, with a rather awkward arthritic gait, and led me into the living room. Here she introduced me to her friend Edna Grant who was sat quietly, in an armchair, in the corner. I don't know what she smelt of because the moth balls were so over powering.

"We've been friends for years haven't we Edna?" asserted Mrs Wattleworth.

"Yesssss", came the restrained response from a rather timid Edna accentuated with exaggerated s's from her ill-fitting dentures.

"Please sit down and take the weight of your feet," invited Mrs Wattleworth pointing to the sofa as she sat herself down in the other armchair.

Mrs Wattleworth was wearing a floral patterned dress. Not only that, the wallpaper was of a very similar floral design as well, and so were the nylon stretch covers on the sofa and chairs.

Her dress was adorned with large pale blue pom-pom type flowers that I think were hydrangeas, and if that was the case, then the flowers on the wallpaper and furniture covers were also hydrangeas.

I was concerned that she was going to blend into the background and I wouldn't be able to see her. However, her snow white hair was the saving grace.

Wouldn't it have been funny if her name was Grace? No? Please yourself.

She also had a ragged piece of fabric sticking plaster across her furrowed forehead suggesting that it had been cut from a roll and had not been individually wrapped. It was so taught over the large lump that it looked as though its contents were fighting to escape and the tell-tale, literally black and blue, bruising of a recent trauma was peeking out menacingly from underneath the edges.

She had clearly been in the wars and quite recently too.

Mrs Wattleworth put Elvis down – on the floor – and then came over to me and started to sniff around my ankles. Elvis that was, not Mrs Wattleworth.

It must have been sheer bliss for her to have a new, different, and dare I say exciting, smell to explore instead of just moth balls or whatever Edna smelt of.

Satisfied after her short foray, Elvis returned to Mrs Wattleworth and curled up on the floor beside her mistress, fell asleep and then started to snore softly and contentedly.

"So are you a big Elvis fan?" I enquired with my rather obvious opening gambit.

"I can't stand him." She looked disgusted at the mere suggestion. "I much prefer Frank…"

I could see she was racking her brain for the name.

Sinatra? I wondered.

"Frank…"

"Frankie Laine", I blurted out. He was definitely her era. I don't know if she didn't hear me or was just ignoring me – not the first person to do that – but there was no response.

"Frank…"

Frankie Vaughn was my next, if pretty poor, guess, but I kept that one to myself.

"Frank…"

Frankie Valli? I was rapidly running out of Franks – that perennial old problem.

"Frankie Goes to Hollywood", she finally declared with evident relief.

I'd completely lost the plot, if in fact there was one.

Edna just sat there nodding like a nodding dog in the back window of a car.

"I love him."

Him?

"Really?" I asked, totally perplexed.

She must have got them mixed up with someone else.

"Oh yes, I love that song where he tells everyone to have a rest."

"*Relax?*"

"That's the one."

She obviously had no idea what it was really about – bless.

Which only goes to prove what Mark Twain said is indeed true. The "Truth is stranger than fiction."

In that case, why not call the dog Frankie?

No I didn't ask. Would you have?

At this point in the proceedings we were rather abruptly, but obviously out of necessity, interrupted when Mrs Wattleworth broke wind. Not just a dainty passing of gas, but flatulence that; (a) sounded like someone blowing an extended raspberry; (b) startled poor Elvis from her slumbers with one hell of a fright and; (c) any pre-pubescent boy would have been proud of.

Who am I kidding; I'd have been proud of it.

Mrs Wattleworth glanced round the room furtively to see if anyone had noticed – it would have been pretty hard not to – then proceeded to blame the dog. "Naughty Elvis", she scolded, "We don't do bottom burps when we have company", and then tapped her on the rump, albeit gently.

The dog looked at me totally forlorn as if to say, 'it wasn't me – honest', and then scampered off into the kitchen whimpering softly. Moments later I could see her through the patio window, in the garden, running round in circles chasing her tail.

Elvis had left the building

The dog was plainly not the perpetrator of the aforementioned 'flatus expelled through the anus' – a mixture of nitrogen, carbon dioxide, oxygen, methane and hydrogen sulphide. Quite simply; a dog that size wouldn't have had the internal volumetric capacity to produce so much gas, unless it had inflated to the size of a zeppelin. And besides, I'm sure dogs wallow in their own stench, they don't run away from it.

When the toxic cloud eventually enveloped me, dense like a 1960's city smog, I was desperate to go and join Elvis out in the garden.

It smelt foul, in fact not that dissimilar to a poultry slaughter house, and I was sure it was starting to make all the hydrangeas wilt. It easily overpowered the moth balls, which was saying something.

I mean, if we already have incontinence pads and sanitary towels, why hasn't someone developed carbon filters for underwear to alleviate such problems? Answer me that.

Edna said nothing; she just sat there perfectly still with a glazed expression on her face. I wondered if she'd actually been asphyxiated.

I used my, not nearly large enough, notebook to try and waft away the smell as though I was an 18th century French courtesan coyly fanning themselves in the corridors of Versailles.

"Can you tell me what happened with the toothpick and the whisky?" I asked, after turning a not very attractive shade of puce and then gasping to take a sharp intake of breath when I thought it was relatively safe to do so.

I could see that she was lost in her thoughts for a moment or two and then she started to tell her tale.

"I poured myself a little drinky poo", she explained, "I usually have a little snifter before I go to bed."

"Every night?" I enquired, hopefully not sounding too accusing.

"Oh yes", was her incredulous reply as if asking me why it shouldn't be every night.

I obviously did sound accusing.

She then stopped and looked straight through me as though she'd finished what she had to say.

"Then what happened?" I encouraged.

"The next thing I knew there was a bit of plastic stuck in my throat."

"How did you manage to get it out?"

She looked confused.

"You coughed it up, remember?" interrupted Edna with her false teeth still clacking in her mouth when she spoke. So she was still alive then.

I smiled reassuringly to acknowledge her contribution.

"Did you witness it then Mrs Grant?" I was hoping she could corroborate the story.

She looked shocked that I'd asked her a question.

"No, no, I wasssn't here", she said hesitantly, "That'sss what Ethel told me."

"When did she tell you?"

"Yesssterday."

"Was it you who suggested she contact the Environmental Health department then?"

"Yessss."

"So how did it get in your mouth if you were sipping the drink?" I turned my attention back to Mrs Wattleworth.

It wasn't that I didn't believe her I just needed to establish all the facts.

"Oh I don't sip it. I knock it back in one go," was her proud response as though there was only one way to drink the stuff.

I could see how the tooth pick could have got in her mouth now.

"When was that?"

"When was what?"

"The incident with the whiskey."

"What incident with the whiskey?"

"When the toothpick got stuck in your throat."

"Oh you know about that then?" was her confused reply.

"Yes you've just told me about it."

She looked at me disbelievingly.

Eventually she replied, "I think it was about a week ago."

"I see you've hurt your head as well", I said stating the bleeding obvious – in which I have a degree. "What happened?"

"I walked into the patio door."

Easily done!

"When did that happen?"

"I think it was about a week ago."

By the look of the bruising, it was a much more recent event.

I was now worried that she was confused as a result of the head injury or perhaps she may have even had a touch of dementia.

"Did you go to the doctors?"

"No I put the plaster on myself."

Ask the wrong question and you don't get the answer you're looking for.

"Sorry, I meant about your throat."

"Yes."

"What did the doctor say?"

"It will heal on its own."

"Have you still got the toothpick?"

"Not in my throat."

"Did you keep it when you coughed it up?"

"Yes" she said and reached over to pick up a small wooden box from the cubbyhole beside her. "It's here, in my jewellery box."

The box she handed me had the most beautiful and intricate marquetry inlay of the 'Rose of Lancaster' decorating the top of it. When I opened it a tiny ballerina in a pink tutu popped up and started to rotate slowly to the laboured refrains of Tchaikovsky's *Dance of the Sugar Plum Fairy*. Inside it was a neatly folded pastel blue paper tissue and nothing else. When I unwrapped the tissue there was the clear plastic toothpick in all its, see through, glory.

"Have you still got the bottle?"

"It's in the kitchen."

"Is it okay if I go and get it?"

"Of course it is."

In the kitchen there were two bottles of whiskey. One was completely empty and the other had a couple of slugs missing from it. Fortunately they were both the same brand, because that was the next bit of information I needed.

"Do you always drink this brand?" I enquired on my return.

"Have done since I was a little girl", she said with conviction. "Did I tell you about the time I got a piece of plastic stuck in my throat?"

"Yes you did."

Even so, she told me again, but at least the stories were consistent.

It was difficult to know if the tooth pick had ever been in the bottle of whiskey at all. Having said that, I couldn't understand why anyone would want to damage the lining of their throat deliberately, and she didn't have any other toothpicks in the house, clear or otherwise.

I explained that I wouldn't be able to get any compensation for her, even so, I would be writing to the distillery, purely from a public health point of view, to see if there were any problems there that they were aware of.

I asked for permission to release her name and address so that she could be contacted by the distillery directly when they had investigated her complaint, and she agreed. So I thought, naively, that that would be the end of that. (That's three 'that's' in one sentence – without even trying).

A few weeks later I received a charming letter from the distillery in question, expressing their deep concern over the matter. They also thanked me for taking the time to contact them to highlight the problem.

It transpired that mine was not the only complaint about a tooth pick in one of their bottles of whiskey; they had received quite a few more. Following an internal investigation it emerged that it was one of their employees who was deliberately sabotaging the product. They also informed me that they had written to Mrs Wattleworth separately to explain what had happened.

So it was all done and dusted.

A week later I received a message from Mrs Wattleworth, would I visit her urgently please?

When I called round to see her she thanked me profusely for dealing with her complaint so promptly and efficiently and I acknowledged her gratitude with, "I was just doing my job."

She then showed me what the distillery had sent her as compensation for her trouble – a case of whiskey.

She had informed me at our original meeting that she only had a wee dram each night. "At my age", she laughed, "This will still be here when I'm dead and buried."

I laughed politely. I wasn't sure if that was actually funny or just plain morbid.

"Please take a bottle with you for all your troubles", she implored.

"Much as I would love to, I'm afraid I can't", I replied.

"Why, don't you like whiskey?"

"Oh yes, I like whiskey," *but not that particular brand.* What a snob, even though I know I shouldn't have looked a gift horse in the mouth – in case it bit. "If I take a bottle it may be misconstrued as a bribe," I informed her.

"Poppycock" was her immediate response.

I always thought that on a poppy it was called a stamen.

After much deliberation I eventually left without a bottle, but I could tell that she wasn't really happy about it.

Perhaps it would have been easier to say that I didn't like whiskey.

Chapter Seventeen

Not long after I started at Northford Council the boss, in his infinite wisdom – actually that's not true, it did have its limits – thought it would be a good idea if I gained a teaching qualification and so sent me off to college to get one.

Hopefully this time the experience will be better than on the last two occasions. And it was. So it would appear that third time is indeed lucky.

He wanted me to be the specialist in food hygiene training for the department, as the other officers already had specialisms. Little did I realise at the time that his idea would stand me in very good stead when I applied for my next job at Borntown.

My first teaching practice was held at the college and, I have to admit, was an unmitigated disaster – and that's putting it mildly. I was told to choose a subject that I was interested in and could talk about for an hour and the college would allocate me a class to teach it to. I chose pest control and they gave me a class of late (not dead) teenage girls who were very prim and proper and studying fashion. Not exactly the perfect match, or even 'Paris Match' for that matter.

What was the college thinking? Why would a group of eighteen and nineteen year old girls be interested in the comparative size and shape of rodent droppings? (My specialist Mastermind subject of choice – only kidding). They were more interested in ratine than rats.

My next teaching practice, which was held outside the college, was generally much better. My tutor came to observe me teach a one day – 6 hours – basic (now level 2) food hygiene course at Northford Council offices. I thought, and hoped actually, he would only stay for an hour or so – wrong – to my dismay he stayed all flippin' day.

Has he got nothing better to do?

At the end of the ordeal, sorry day, he informed me that he thought my performance was pretty good considering and he'd actually learnt quite a lot about food hygiene into the bargain.

"Considering what?" I enquired.

He went on to tell me, quite sensitively, that he might have learnt even more if only he could've heard everything I said. It would appear that at times my voice was drowned out by the jangling of keys. Or more precisely, by me jangling the keys in my trouser pocket – an annoying habit I had when I was nervous. I stopped leaving keys in my trouser pockets after that.

During my, nearly six and a half, years at Northford Council I taught food hygiene to countless different businesses and organisations including a group of YTS (Youth Opportunity Scheme) girls. I mention this group specifically because they taught me another valuable lesson.

I'd spent the first half of the morning, of their course, rabbiting on about bacteria – where they could be found and how they multiplied – to a sea, nay, ocean, of blank faces.

What am I doing wrong? Why do they look so vacant?

At the coffee break (other beverages were available) I found out. One brave (but not a Native American warrior) girl sidled up to me and asked, "Ere mista, you keep talkin' 'bout bacteria, what are they then?" I think she might have been cockney, she was certainly cocky.

"You know," I said, but she obviously didn't or she wouldn't have been asking the question, "Germs, bugs, things that make you ill."

"Ooooooh," she said, sounding somewhat relived, and then turned to her mates and declared, loudly and proudly, "E's talkin' 'bout germs."

That single, short, sentence was greeted by a chorus of approving 'Aaaaaaahs' having obviously made considerably more sense than the multiple, long, sentences I'd been spouting forth up until that point.

And the lesson I learned that day; never assume people understand the vocabulary you're using.

She then went on to ask, "So 'ow'd you know all this stuff then? You a know it all?"

"No," I said modestly, because I wasn't. "It's my job."

"So are you an Elf Inspector then?"

"Yes," I said proudly, even though I'd never been called that before, but I have to admit I found the term kind of cute (soppy sod).

Back in the day, the test at the end of the course consisted of twenty short questions that each required, up to, a two line written answer. None of that multiple choice malarkey you get nowadays, but I fully understood why it had to change – especially after this next experience.

I always used to mark the papers straight after everyone had finished the test – until they eventually had to be sent to the CIEH to be marked – so that everyone knew their result before they went home. There was only a maximum of twenty candidates on each course, so it didn't take too long – especially because I knew the answers. I was motoring along, quite happily, through the papers at a fair rate of knots when I suddenly came across one I couldn't read. Not because the handwriting was poor or it was written in a language other than English, but because it looked like complete and utter gobbledegook to me. I stared and stared at the paper, and then stared some more, to see if I could make any sense of what I was staring at. Then all of a sudden I realised that one of the words was an anagram. In fact when I looked a bit harder I could see that most of the words had been written as anagrams. It was at that point I surmised the candidate probably had some form of learning difficulty, and more than likely dyslexia.

Now I always used to say, at the start of my courses, that if anyone didn't want to take the written test, for whatever reason – I didn't need to know the reason why – then I would arrange an oral test for them. I had a blind Bambi (no eye deer – no idea, please yourself) why this person hadn't informed me, unless of course they were too embarrassed to do so, but when I finally finished marking their paper they'd scored full marks (with no head lice to be seen) proving that difficulties don't have to be barriers.

Following the popularity and success of the basic food hygiene courses I went on to teach the intermediate (now level 3) and advanced (now level 4) courses as well, which would also benefit me when applying for my next job.

Chapter Eighteen

Shit happens!

But of course you don't need me to tell you that, you already know. Still, if it's any consolation – and I very much doubt it is – it happens to other people too.

It was certainly happening by the bucket load (if that's how you measure it – or would it be better to weigh it? That's fine if it's solid, but useless if it's liquid – it would have to be by volume then) in the sleepy village of Upton Abbot on that particular Monday morning.

Why?

Because there was a major food poisoning outbreak breaking out, that was why.

It was at times like that I wished I'd had shares in toilet roll, antiseptic healing cream and air freshener companies – Ka-ching.

The villain of the piece – boo hiss, it's behind you (literally) – was a particularly nasty organism (that's with a 'ni') that causes misery, suffering – and a really sore bottom - whenever and wherever it attacks. Allow me to let it introduce itself; 'The name's Bacter, Campylo Bacter.' Now that doesn't mean a vertically challenged (short in old money) effeminate bacteria, what it does mean is a 'twisted bacteria' – and if that means twisted as in sick, then it's very aptly named. It has a particularly nasty streak and is capable of leaving a particularly nasty streak too – in the toilet bowl mainly, but sometimes in underwear as well.

'I'll have you shaken and stirred,' said Bacter. Just thought I'd continue the Bond reference a bit longer.

By the way, it can't actually talk.

Now some of you might have heard of Campylobacter. Worse, you might have suffered at its hands – even though it doesn't have any hands only flagella (as in Benjamin). After all, it is now recognised as one of the most common causes of bacterial foodborne disease in the U.K. and many other developed countries in the world for that matter. I'm sure it's prevalent in underdeveloped countries as well, but they probably don't have the facilities to isolate and report it.

Its symptoms, in case you were wondering, include nausea, diarrhoea (explosive and bloody) and worst of all the diarrhoea erupts – like an exploding Vesuvius from the rear end – without any warning at all. (I'll tell you about my close encounter of the turd kind later on – bet you can't wait). Oh, and sometimes there's vomiting and the victims may also get stomach cramps and a fever. All in all, it's a nasty little bug-ger! Although it could be very useful for fast weight loss, but I wouldn't recommend it.

The incubation period – that's the time between eating any contaminated food and the symptoms starting – is usually between 2 – 5 days, although in some cases it may take as long as 10. The symptoms can last up to a week but in most cases the sufferer is better within 3 days. And that's the strange thing about it, the symptoms stop as abruptly as they start.

What's that all about then?

Campy (as it's affectionately known) can be found in raw poultry – that's why you should cook it thoroughly and wash your hands after touching it – unpasteurised milk (not easy to purchase these days) and some pets, including cats and dogs, may also carry it.

I hope you're taking all this in because there'll be a test at the end.

Following a wedding reception on the Saturday, the Monday morning saw a fair few notifications of suspected food poisoning – considerably more than would have been deemed usual anyway. The following and subsequent days, even more reports came pouring in as many of the guests at the aforementioned nuptials were presenting at their GPs with a variety of symptoms that included the notorious double act: diarrhoea and vomiting.

During my initial investigations I established that the reception had been held at a pub on my patch. Some of the guests had gone back home out of our area and so the EHOs in those authorities would have to carry out those investigations for me.

I had the classic situation of a foodborne-disease outbreak on my hands, its definition being 'an incident in which two or more persons experience a similar illness from the ingestion of a common food, and epidemiologic analysis implicates the food as the source of the illness' [sic].

When I went to investigate the first batch of notifications I asked the victims where they'd eaten recently and they all said they'd been to a wedding reception the previous Saturday. I then asked them if they could remember what was on the menu and what they'd actually eaten. Everyone told me that it was a buffet, but most of them couldn't remember the whole menu, however, between them I managed to glean what the majority of the items were. Fortunately though they could all remember what they had eaten individually. There were chicken drumsticks, sausage rolls, prawn vol au vents, ham sandwiches, potato salad and the coleslaw. For dessert there was Black Forest Gateau and another type of cream cake that no one knew the name of.

All I was really concerned about was that it had cream in it and could therefore potentially support bacterial growth.

The very next day (Tuesday) the shit really hit the fan, even though there wasn't a fan in sight, and we were inundated with notifications to the point I had to recruit the services of my colleagues to help me out.

When we reached the outskirts of the town there was an all pervading smell of manure filling the air.

Surely they haven't blocked the drains with the excessive amount of poo they're producing?

No, the local farmer was sludge spreading.

It has been well known, for eons, that a common practice in rural areas is to spread animal manure on farmland to act as a fertiliser. This is a practice that is now generally referred to as organic farming, i.e. not using any agro-chemicals. What is probably less well know is that in areas where there are sewage treatment plants the treated 'human manure' is often spread on the fields too. This is known as sludge spreading.

The majority of the residents in the area accepted the fact that manure was used on the fields, but some would regularly complain when the farmers were doing it. Why move to a rural area if you don't like rural smells, or noises for that matter? And that reminds me of another story.

Not long after I started at Northford Council, I received a complaint from a nurse who'd moved away from the hustle and bustle of one of the major midland cities to the serenity of the rural village of Rosslea. She regularly worked nights at the local cottage hospital and so slept during the day. She rang me to complain about the cows in the field next to her cottage "mooing" during the day and keeping her awake.

When I visited her at her cottage, Miss Smith (don't think it was a pseudonym) was dressed in her sister's uniform – not her sibling's uniform, but that of a nurse who has been promoted to sister. I wasn't sure if she was wearing it because she had just got back from work, was about to go to work or to prove that she was actually a nurse. Anyway the cows were mooing when I arrived.

"What are you going to do about it?" were her first words when she opened the stable type front door. Not even a hello.

"What would you like me to do about it?" I asked, probably sarcastically. I couldn't believe that someone was actually complaining about cows mooing. I was half-expecting someone to jump out and say, 'smile you're on candid camera.'

"Tell them to stop," was her very serious and straight-faced request.

I was tempted to ask her if she'd tried telling them to stop and if she had and they'd ignored her request, why did she think that I would have any better luck.

"The problem is," I reliably informed her, "I don't speak cow." Yes I actually said those exact words – flippant of me I know. "I can't even serve a notice on them (under The Prevention of Mooing Regulations 1848) because I can't write cow and they probably can't read anyway," I continued. *And I would be laughed out of court if I tried to enforce it.*

I bet she used to sleep like a log through the endless traffic noise etc. in the big city, but she obviously couldn't get used to the peace and quiet of her new life.

When she realised that I wouldn't/couldn't do anything about the situation she abruptly said, "Goodbye," and then slammed the door – not quite in my face, but as good as.

And a fond farewell to you too.

Anyway back to the sludge spreading.

Fair enough, there were guidelines which stated that the spreading should be done when the wind was blowing away from domestic properties so that the smell was wafted away. They also stated that rain should be imminent so that the smell was dampened down. Inevitably these guidelines were never adhered to, hence the complaints. The problem was, once the manure was on the field there was no way of taking it off again and besides the smell didn't last that long.

The time when most of the residents complained was when the farmer was sludge spreading. Again, the guidelines were rarely observed and I have to admit that sludge does not have the same sweet rustic smell of animal manure. However, the main reason for the complaints was that the sludge came from the new town on the other side of the river.

Now that was partly true. The sewage farm was on the banks of the river on the Rosslea side and not only treated their waste, but also that from the other side of the river i.e. the new town.

I mean how petty can you get? Shit is shit isn't it? If I'd told them that the farmer was only using shit from their side of the river would they have accepted the practice? Probably! It was all very parochial.

Anyway, while my colleagues were busy interviewing the unfortunate victims I went to the incriminated pub to see what more I could find out.

The owners weren't there when I arrived, but the 'chef', a Mr Lamb, was. I explained the whole sorry situation and asked him if he would answer a few questions about the Wedding Reception for me. He seemed a bit reticent to at first, you might say even sheepish, but in the end he did provide me with some very useful information, although I don't think it was necessarily intentional.

"Were you working on Saturday?" was my first question, because if he hadn't been then there would have been little point in asking him anything else. He informed me that he was and then went on to tell me that he'd been employed from an agency 'temporarily' on the Friday, but actually ended up being in charge because most of the other staff didn't turn up on the Saturday.

On the day in question he'd arrived at 06.00, as he was told to, so that preparation could start early and everything would be ready in good time. I asked what other help he had and he said a kitchen porter came in at 09.00 to do the washing up. This was not looking good! I asked him what the menu consisted of even though I had a pretty good idea from the people I'd already interviewed. He told me there'd been chicken drumsticks, prawn vol-au-vents, sausage rolls, ham sandwiches, potato salad, and coleslaw as well as two different types of gateaux – so at least that all tallied up.

I then went on to ask him how he'd prepared everything. He told me that the chicken drumsticks had been thawed in a mixing bowl, at room temperature, overnight to make sure that they were completely defrosted. Later in the conversation he told me that the same bowl had been used to make the potato salad in. When I asked if it had been washed in between he said, without even flinching, he'd wiped it out with a damp cloth. The same damp cloth was later used, and I quote, "To wipe over all the serving dishes because they were a bit dusty. Yummy, yummy, yum, yum, food poisoning for my tum, tum!"

He also told me that because there wasn't enough fridge space all the prepared food was laid out, with the exception of the gateaux, in the dining room. And so it sat about on a warm summer's day for about five hours – give or take a million bacteria or so – until the wedding party arrived, late of course – bloody photographers. He said he'd kept the windows open though to keep the food nice and cool – and let the flies in as well no doubt, and we all know what happens when a fly lands on your food.

You don't?

Flies don't have teeth and are therefore unable to eat solid food. What they do have is a straw-like mouth part called a proboscis. To eat they have to vomit up their previous meal, including digestive juices, onto food and then perform a jig to help liquefy it, whilst stamping in a few (million) germs as well. When it's all nice and runny they suck it back up again, leave the inevitable calling card, fly off, and then it's your turn to eat.

Well you do now.

I asked him if he had a basic food hygiene certificate. His response was to just stare at me as though I was speaking a foreign language. After a long time he asked me what that was. I must admit, I was beginning to wonder myself.

Finally I went on to do a full food hygiene inspection of the premises because, according to the file, they'd not been inspected for over eighteen months when there had only been a couple of minor contraventions.

The kitchen was small. Way too small to be catering for large functions. But I bet the profit margin was huge.

It was blatantly obvious that the place had gone downhill. In fact it was so far down the hill that it had crossed the valley and was starting to climb the hill on the opposite side.

Just as I was nearing the end of the inspection the owners returned. I explained who I was and why I was there at which point they went absolutely ballistic. Firstly they had a go at me for suggesting that they'd caused the food poisoning outbreak and then threatened me, albeit indirectly, that they knew the Kray twins. I did want to say that I knew of the Walton sextuplets, but I got the distinct impression it wasn't a game of one-upmanship. They then went for the chef, big time, for having told me everything.

I reckoned Mr Lamb was for the chop when I left.

In the end sixty seven out of the eighty two guests were positive for *Campylobacter jejuni* (yes it has a second name as well, in fact there are over thirty different types of it). Even the bride went down (steady) with it.

When I presented the case to our solicitor, who wasn't a council employee, he said that he didn't think there was enough evidence to prove the food had been 'unfit for human consumption' and to prosecute for food hygiene offences only.

How much evidence did he think we needed?

It transpired that one of the company directors (who owned the pub) was a barrister (that's a barrister and not a barista) and was going to defend them in court, so we in turn, at huge expense I might add, had to appoint our own barrister, even though it was only going to be heard in the magistrates' court.

The first meeting I had with our barrister was on the morning of the trial (very Rumpole of the Bailey) – nowt like leaving it to the last minute. The first thing he said to me after, "Hello pleased to meet you," was, "Why aren't you prosecuting for 'unfit food' because there's enough evidence?"

Why indeed. I reckoned our solicitor just couldn't be bothered with what he saw as a more complicated case – lazy sod.

There were three elderly magistrates hearing the case and they were all male. I didn't know if that was because there weren't any female magistrates or because it just wasn't their turn on the roster, but it certainly looked as though these three had drawn the short straw. And they made no bones about the fact they didn't really want to be there because they were trapped indoors on such a gorgeous summer's day.

Didn't I mention the weather? Sorry. Well it was.

I was positive that they'd have rather been lounging around in their respective gardens with an ice cold G&T in one hand and a good book in the other. Even an average book would have been better than what they were about to endure.

One of the gentlemen – and all three of them were most definitely gentlemen – was wearing a toupee, and not a very good one either. It wasn't even the same colour as his own hair which was making an unwelcome appearance (like a pig at a bar mitzvah) from the sides and back of it. It wasn't even stuck down properly a kept lifting in the breeze from the open window. It also looked as though it had had a home perm applied to it. Perhaps he wanted it to look like a barrister's wig.

I was incredibly nervous as I entered the witness box as this was my debut giving evidence. I was indeed a virgin persecutor, sorry prosecutor – and that isn't a Richard Branson company. Throughout the ordeal I was desperately clenching my cheeks (bum) together in a desperate attempt to prevent my external anal sphincter from relaxing. It didn't.

After we'd presented the case for the prosecution it was the turn of their barrister to cross exam me and the first thing he did was attempt to discredit me with a character assassination.

"Who are you to say that these premises do not comply with the food hygiene regulations?"

Are you kidding me, you know precisely who I am.

"I'm a qualified EHO."
"What's an EHO when it's at home?"

So you're going to be pedantic are you?

"An environmental health officer."
"And just how long have you been an E-H-O?"
"Just over two years."
"Is that all?"
"Yes."
"And do you really think that's long enough to prove that you are competent to inspect such premises and make these spurious accusations?"

I just wanted to curl up and die. This wasn't supposed to be about me; it was supposed to be about the food hygiene offences. I didn't know what to say. I was hoping that our barrister would object to the line of questioning because it was irrelevant to the case. But he didn't. When I asked him why later on he told me, "You were doing just fine. If anything it showed the defence were clutching at straws."

I spent all morning being grilled (I am a bit tough, perhaps braising would have been better) by the defence and at the end of the session the magistrates decided that they wanted to visit the pub so they could visualise what was being talked about. Like the contraventions would still be there. The owners would have obviously sorted the place out by now. I was worried that the visit might weaken our case, but there was nothing I could do about it.

Looking back, I think they just wanted to get out of the courtroom for a bit to be honest.

After lunch, which I couldn't face, we all piled into a variety of cars, Uncle Tom Cobley – although I don't know why he had to go – and all and went to the pub, which fortunately was only a short drive away.

I'm not sure (well I am really) if the magistrates actually knew what they were looking for, or at for that matter as they ambled round the kitchen, but they probably thought they looked very professional – they didn't.

The case went on much longer than everyone had anticipated. It should have only realistically lasted a day, but it just seemed to drag on without any end in sight. When the case for the prosecution was finally finished it was decided to call it a day. At this point it was plain to see that the magistrates were totally miffed because not only had they missed one gloriously sunny day stuck indoors, they were going to have to miss another.

I went to collect the court diary from reception to book it for another 'fun-filled' day and as I was walking through the ante room I heard the magistrates talking.

"They are obviously guilty."

"No doubt about it."

I agreed entirely, that's why I was taking the case, but I thought that they should have at least heard the case for the defence before making their decision.

After a second, mind-numbingly boring, day in court the defendants were indeed found guilty (who would have guessed?) and were given a substantial fine. They later appealed the severity of the fine only to have it increased – which served them bloody well right.

Now I did promise you that I would tell you the story about me and Campylobacter – makes it sound like my best friend, it isn't – so here it is.

Sad to say, I have suffered from the little blighter in the past even though I didn't know it at the time. This was because I was just a teenager and not at all interested in the different types of food poisoning – funny that.

It all happened on a warm summer's evening on a journey back from Manford Town Hall. My dad had taken me there to be presented, by the Mayor no less, with the third prize, in an under sixteen's art competition.

This was funny really, because I can't draw for toffee – or any other sugary confection for that matter. And what is even funnier is that I got a grade 2 in my art 'O' level – equivalent to an A at GCSE.

Anyway, on the bus home my stomach started to make some very strange gurgling noises, not unlike air trapped in a central heating system. Not only were the noises strange, but they were embarrassingly loud as well, to the point people were starting to stare at me quizzically, including my dad.

"Are you alright son?" he enquired, obviously concerned for my wellbeing.

"I'm not really sure," I replied over the continuing, and getting louder by the second, gurgling sounds.

I knew that the next bus stop had some public toilets close by and so I told my dad that I needed to get off the bus pronto. I thought I just wanted to break wind, but was worried it might be more than that.

I managed to get to the toilet just before I was touching cloth. As I plonked my arse on the, not too pleasant – but needs must in an emergency – toilet pan I evacuated my bowels with an almighty splatter that lasted for several minutes, but seemed much longer. It was gastrointestinal Armageddon down there. I could have pebble dashed the Great Wall of China and still have had some left over for the Berlin Wall. I had been only seconds away from literally shitting myself – pardon my French – and the toilet looked even more the worse for wear by the time I'd finished with it. It took more than one flush to dispose of its faecal contents. When I eventually got round to cleaning myself up I did notice that there was some blood on the toilet paper, well the tissues I used from my pocket because as usual there was no toilet paper in the cubicle. I assumed it was because some of the superficial blood vessels around my anus had burst under the extreme pressure of the onslaught and so didn't say anything about it to my dad for fear of worrying him and then forgot all about the incident until I was studying Campylobacter at college.

At least when I talked about Campylobacter on the food hygiene courses, I could say that I was speaking from experience. And what an experience it was - hopefully never to be repeated - and not one for the 'bucket list' either.

Test Question:

Which one of the following is a symptom of food poisoning?
A Diarrhoea
B Gonorrhoea
C Glue ear

Chapter Nineteen

The hills are alive with the 'Sound of Mutilation', or, there's 'Death in Them There Hills'.

One of the more out of the ordinary jobs I was involved with was as an undercover agent. Not quite James Bond at the Secret Service, more Basildon Bond at the Self Service.

The EHOs of a neighbouring local authority were convinced that lambs were being slaughtered illegally on a farm, on their side of the boundary, in the hills that separated the two authorities. The best way I can describe its location is that it was in the middle of nowhere, miles away from anywhere. They suspected that the meat was being supplied to a butcher's shop in their main market town and that there was no way of knowing if it was fit for human consumption, or not as the case may be, because it obviously wasn't being inspected. They couldn't even check if the meat had been stamped once it had been butchered into the various retail joints.

This, of course, had potentially serious consequences. If it was the case and unfit meat was getting into the human food chain then the health implications could have been catastrophic. As we have seen over recent years.

Because the officers knew that the perpetrators of this heinous crime (very CSI Northford – I don't think) would recognise their cars they asked us if we would carry out the undercover surveillance for them.

Of course, but why did it have to be in February?

On the third consecutive night of the stake-out it was my colleague's, Joe, and my turn to do the honours. And I really do believe that we should have been awarded at least an OBE – for Other Buggers Efforts.

As we pulled up on the remotest part of the hillside that looked down over the farm it was initially a beautiful evening – well late afternoon really. The stratocumulus clouds that covered the sky were backlit by a fiery sunset that glowed like the dying embers of a fire. The colours gradually mutated from an intense yellow to a warm orange and then finally a sumptuously deep purple (and I'm sure that if there'd have been water there, it would have had smoke on it too). It looked as though the sky was actually aflame. Sadly though, like all things that burn so brightly, it burned itself out all too quickly and when the sun finally sank behind the now silhouetted hillside the temperature dropped like a stone. This was then followed, in quick succession, by a host of moonlit clouds that raced across the heavens like ghosts in an afterlife marathon, all following each other closely for fear of lagging behind.

It was then that my attention was grabbed by the largest cat I've ever seen in my life – and yes it was only a domestic cat, not an escaped big cat from the local zoo – as it stalked across the ground, as if it owned it, in front of my car. It crouched down low and then slunk along menacingly, every inch the hunter, creeping ever closer to its, not much longer for this world, prey. That's the prey I couldn't see with my incredibly poor night vision, but it obviously could. Then, without any warning, the cat pounced landing firmly on the unsuspecting creature – probably a mouse – with an almighty thud. One swipe with its size 9 paw, a lethal grab of its vice-like jaw and it was all over. (I knew it was all over because the fat lady had started to sing). The cat then padded off triumphantly towards the bushes with its kill hanging limply from its mouth.

Now my VW Polo was a comfortable enough vehicle for driving from A to B in – and even from B to A for that matter – but not for being cooped up in, for hours on end, in the cold with no light to read by, no radio to listen to and no toilet to – well you know what in. And we all know what cold weather can do to the bladder.

Bring back the caravan, all is forgiven.

By now the clouds had completely covered the sky, which had become starless and bible black, so to keep ourselves entertained we decided to play some games, including 'I spy'.

Joe: "I spy with my little eye something beginning with N."

Me: "Night."

Joe: "Correct, your turn."

Me; "I spy with my little eye something beginning with D."

Joe: "Darkness."

Me: "Correct."

Joe: "My turn again. I spy etc...... B."

Me: "Blackness."

And so it went on, but not for much longer obviously.

161

It seemed an interminably long night as we tried to apprehend the culprits in between alternate bouts of trying to acquire some sleep. Unfortunately, we failed to capture either.

The following day the birds started chirping as soon as dawn cracked – she was useless under pressure. The morning sun rose from its slumbers slowly yet majestically whilst I fought with extreme tiredness – we never really got on – and at that time of the year it cast a long and sinister shadow of the hills, across the dew soaked grass, which resembled a single accusatory finger pointing directly at us.

I must have fallen asleep at some point because I was woken, not by the birds, but by condensation dripping off the ceiling of the car onto my head. I was now cold and damp. I eventually clambered, dithering, precariously out of the car and started to unravel myself like a butterfly emerging from its chrysalis. As the sun bathed my face in its weak, but still warming, rays, I tried to establish contact with my limbs and get them functional once again.

Needless to say we didn't catch anyone slaughtering animals illegally on the farm. It was just another wild goose chase. (Never quite sure if that means you are chasing wild geese, or wild geese – as opposed to tame ones – are chasing you, because as you know I have been chased by tame ones). As with a lot of environmental health work, it was a complete waste of time. The number of complaints any department receives each year that cannot be justified is ridiculous. Unfortunately some people have nothing better to do than complain, and more often than not, about nothing. In this particular case, either 'the gang' had sussed us out and knew what we were up to, or it just hadn't been going on in the first place. (I don't think there are any other options). However, it was an experience – of sorts – but one I never wanted to repeat again, and luckily never did.

Chapter Twenty

I can't abide cockroaches; in fact I'd go so far as to say that I detest them. (I don't like the word hate, it's used far too often and there's too much of it about).

To put it quite simply, cockroaches are vile creatures. VILE, **VILE**, **VILE** I tell you – with my left eye twitching manically and my right shoulder pulsating in spasm like some stereotypical megalomaniac. And that's a fact; the naked (cheeky) truth; gospel (according to me anyway), and if anyone wants to contradict me then I will happily challenge them to a duel – aerosol pesticides at dawn. If you don't believe me – and why wouldn't you? – Then read on and I'm sure I'll bring you round to my way of thinking. But I must warn you, you do so at your own peril! Cue the sinister music as the lights dim and flicker and the room starts to feel icy cold.

It was late one Friday afternoon – so called emergencies always happened on a Friday afternoon – when I received a call from the on duty sergeant at the local armed forces base saying that they had a problem with cockroaches. He claimed that although they had a contract with a well know pest control company – which will remain nameless – they were still seeing cockroaches during the day.

Not good. Not good at all.

Now as you may or may not be aware cockroaches live in a wide variety of environments all over the world with the four best known species being; the American; the German; the Asian; and the Oriental.

Cockroaches are quite large insects, most of them being the size of a thumbnail -- without the extension – although the Australian Giant Burrowing Cockroach can reach up to 9cm in length and weigh around 30g.

I wouldn't like to meet one of them on a dark night, or a light one for that matter.

The main problem with them is that they all adapt readily to their surroundings, even though they much prefer the warm conditions found within buildings and especially kitchens.

Because they are predominately nocturnal insects – they will naturally run away when exposed to light – and if they did in fact have some wandering about during daylight hours then there must have been nowhere else for these waifs and strays to go. In short, there must have been so many crammed into the structure of the building that these vagabonds had been kicked out and told to look for alternative accommodation.

Think it's a joke?

Do I look as though I'm laughing?

It's absolutely true.

The sergeant asked – well pleaded to be honest, he was a right little pleader – for us to take over the contract and eradicate the problem once and for all, and like all local authorities; we'd do anything for money – except that – I was like the man from Delmonte and said, "Yes." I did, however, inform him "It's too late to visit today, but I will come out first thing on Monday morning with our pest control officer and see what we can do."

The department's pest control officer was Dave Gracie whose best friend was Jack Daniels (who didn't work in the department). He was also very well acquainted with Johnnie Walker and Jim Beam too.

I've also met all three of them once or twice in the past, but could only take them in small measures or else they would cause me to have the most terrible headache the following morning.

I'd also heard it said that John Smith could be an influence on Dave as well, especially if the others weren't about.

He had a rosy complexion (red that is, yellow would have been something completely different) with numerous burst thread veins on his face, a bulbous nose and breath you could strip wallpaper with, but a very kind heart.

The plan of action when we got to the site was to assess the situation and find out:

(i) What area the infestation covered; this is important to know because the poison would have to be laid way beyond the boundaries of the infestation so that the cockroaches would always be walking into it if they tried to escape.

(ii) Which species of cockroach it was because different species have different life cycles which would influence the treatment.

The two most common types of cockroach in the UK are the German and the Oriental.

I used to ask groups, when I was talking about cockroaches on a food hygiene course, if anyone knew what the difference was between these two species. On one occasion someone put their hand up and said that the German cockroach was the first by the pool in the morning putting its towel on a sun lounger. I did have to inform them that they were wrong.

The actual differences are; the German cockroach – which is probably the most troublesome – is typically about 1.1 to 1.6cm long and varies in colour from tan to almost black (new Farrow and Ball colour), and although it does have wings it can barely fly at all. It is thermostatic – preferring confined spaces – and is particularly associated with restaurants, food processing plants, hotels and nursing homes – to name but four of its favourite haunts. It reproduces faster than any other UK residential cockroach growing from egg to reproductive adult in approximately 123 days or four months. The female carries the egg case, or ootheca, in her abdomen until it has completed its maturation when the baby cockroaches, or nymphs, hatch out over a twenty-four hour period.

On the other hand the Oriental cockroach – also known as the waterbug, because they prefer dark moist places like sewers and drains – is much larger than its German counterpart reaching between 1.8 and 2.9cm in length. It is a dark brown to black colour and has a glossy body. The female has a much wider body than the male – but for goodness sake do not, under any circumstances, tell her that – and the odd male is capable of very short flights ranging from 2 to 3 metres. In this case the ootheca is formed a day after mating and deposited between two and seven days later. The nymphs, however, do not hatch until forty-two days later – which would be a great name for a film.

Oriental cockroaches are notoriously harder to eradicate because although the adults can be killed quite easily, within forty-two days of the eggs being laid the next generation of nymphs will hatch out, grow into cockroaches, and the infestation will continue.

(iii) Roughly how many cockroaches were there in the infestation because this would determine the amount of insecticide that would be needed as well as the amount of time it would take.

We went along suitably armed with an aerosol can – which probably had CFCs in it – of a product know as 'Drione'. Drione was not freely available at the time, but can now be bought on-line. It does kill cockroaches, but also contains a flushing agent that encourages the little blighters to emerge from wherever they're hiding.

The good news was that the kitchens were completely detached from the rest of the site which meant that the infestation was probably contained to just them.

Hallelujah!

The bad news was that they were absolutely gi-nor-mous, the kitchens that was – not the cockroaches. If it had been the cockroaches then you wouldn't have seen me for dust – the small ones were bad enough as far as I was concerned.

Yes, it had all the makings of yet another cheap and tacky horror movie; 'Planet of the Cockroaches', complete with wobbly sets, pathetic special effects and acting so ham you could have served it with egg and chips.

There were two huge/colossal/gigantic – I really don't know which word to use, so you can choose which you prefer – kitchens linked by a wide corridor and then two massive/immense/humongous – ditto – dining rooms on either side of the kitchens that could have easily seated more than five hundred people at one sitting. I'd never seen a catering operation on such a vast scale, and I haven't since.

I hadn't inspected the premises for food hygiene before because it had the privilege of crown immunity, which meant it was immune from prosecution. The crown still has immunity from prosecution under the Food Safety Act 1990 (can you imagine closing the kitchens at Buckingham Palace on the night of a State dinner?), but the MOD lost it on 1st April 1992.

I was genuinely shocked to find that such a prestigious premise was in such a poor state of repair. In fact it was a bit of a dump really. There were loads of cracks in the, not very, white wall tiles, which gave the impression of a mosaic mural – but without the picture. Not too much of a hygiene concern because bacteria aren't capable of parachuting down from them – honest – and there was no evidence of the staff throwing food at the walls either, but an ideal hiding place for cockroaches.

A lot of the floor tiles were also badly cracked and broken, yet another hiding place, although I assumed that if any food fell on the floor it would be thrown away immediately. The five second rule is a myth.

There were also a couple of broken drain covers in the floor which would have given the cockroaches' easy access to the kitchen. A stupid place to have them in the first place if you ask me – which I'm sure you would if you met me – but they should have at least been maintained to help prevent the problem in the first place, because as Desiderius Erasmus once said – he may have actually said it more than once, I don't know – 'prevention is always better than cure'.

Most of the door jambs were coming away from the walls, so there were even more hiding places for cockroaches, and on top of all that, literally, was a suspended ceiling which cockroaches love because the can scurry from one end of the premises to the other, in the dark, and no one would have the slightest idea that they were there – unless of course they were wearing clogs.

On the day we arrived the site had just closed for two weeks for the annual summer holiday, however, a lot of food had been left in the heated trolleys, so the cockroaches were having a bean feast – and it wasn't just pulses on the menu. Not that they needed that food anyway because cockroaches are omnivorous scavengers, i.e. they eat all kinds of food indiscriminately, but prefer meats, starches, sugars and fatty foods. They are also cannibalistic i.e. they will eat their own species, but they don't necessarily wait for each other to die before doing so. Then when they've finished eating, like all 'animals' do, they eventually need to defecate and other cockroaches will come along and eat that too.

I told you I'd bring you round to my way of thinking.

Cockroach 1: "What's for tea tonight?"

Cockroach 2: "Well I ate Uncle Henry yesterday, so it'll probably be his protein that I consumed; digested; the excess of which I've just excreted!"

I sincerely hope you're not eating while you're reading this bit, and I mean that from the heart of my bottom.

Perhaps there should be an airline type sick bag at the back of the book for emergencies.

Well when we started to spray the 'Drione' into all the cracks and gaps that the little beasties could hide in, of which you are now aware were many, it was like turning brown water taps on. They just gushed from every orifice that we sprayed into, and some we didn't.

I had them crawling, well running actually, up my trouser legs – on the inside. They were even dropping down the back of my shirt collar from the suspended ceiling. I was actually afraid to open my mouth and speak for fear of one; or worse, more, wandering in and me swallowing it/them. This reminds me of that old joke: "What's worse than finding a cockroach in a pie… finding half a cockroach." Bum, bum!

I am available for weddings, christenings and bar mitzvahs – and I do give change.

At one stage I was covered in a mass of marauding cockroaches.

So now you can see why I don't like the disgusting sods. As far as I was concerned it was my worst nightmare come true, my Room 101.

Eventually when this 'intrusion' had all died, the whole of the floor was completely covered with dead cockroaches that crunched underfoot – like frozen snow on a bitterly cold but sunny winter's morn – as we walked about to assess the carnage.

When we returned to sweep them up the following day we had a pile so high that it resembled a termite mound. And that was just the initial assessment.

Just as an aside, if you do manage to collect any cockroaches here's a suggestion what to do with them. Put them (live) into a deep bowl, pour cooled melted chocolate over them – as if you're making rice krispie cakes – and then eat them straight way so that they wriggle as they go down.

I'm sure that they will do something similar somewhere in the world.

Pigeons poo – prolifically – period!

Not only is that a true fact, it's also a beautiful piece of alliteration.

Pigeon's poo's positively putrid, perpetuating persistent problems.

I'm on a role now.

The quaint little village of Idsall had a Major Problem (ex army) as well a considerable problem with this excruciating excrement.

I'll stop now.

For starters, there was an abnormal population of the pesky pests, and I don't mean abnormal in the sense that they had two heads or only one foot – some were indeed much taller. What I do mean is there were considerably more of them than you would expect in an area of that size and 'human' population, mainly due to the fact that the 'human' population were regularly feeding them. That's the same 'human' population that were complaining about them – naturally.

The railway bridge that went over the main road into the village had become the main roosting place for the pigeons in the evenings as well as a perch during the day. As a result there was a considerable amount of poo cascading down the bridge like excremental stalactites, but there was even more dripping onto the pavement and road. It wasn't so much a bridge over troubled water – it was more a bridge over, well you can make your own title up.

At its worst the pavement was turned into a skating rink with the slippery slimy sludge and quite a few people did actually fall, some really badly.

Now you couldn't just go around killing pigeons indiscriminately, you needed a licence from the Ministry of Agriculture Fisheries and Food (MAFF), now defunct, which gave permission to use the narcotic that would actually kill the pigeons humanely. The terms of the licence were very specific and this is what we did to comply with it.

The day before the cull we went to Idsall to ascertain where their feeding grounds were first thing. Early the next morning we were there again, just as the sun was rising, but we didn't hear a maiden singing in the valley below – mainly because there wasn't a valley below and I doubted there were any maidens in the village either. We surreptitiously laid the bait on the small recreation ground where we'd established the pigeons liked to feed and they flew down like a flock of hungry gannets that had never seen food before. *This is a doddle* I thought, until an old banger (car, not sausage) went past and back fired. The startled pigeons took off in fright, however many there were, and then started to drop out of the sky – stone cold dead.

Unfortunately we hadn't thought to count them when they were pecking away on the ground, but now we had to account for them all as we scurried through the streets retrieving the carcases, feathers absolutely everywhere, and dashing into people's gardens to rectify the carnage. I didn't think anyone would believe it was foxes.

After that the problem seemed to be solved until a few weeks later I received a call complaining about the pigeons on the railway bridge in Idsall. Another 'kit' of pigeons had established itself in the area mainly because the 'human' population were putting food out for them again.

The last 'pest' story I want to tell you is about a house I was called to where the tenants were complaining that they were regularly being bitten by something. My first thought was that it was probably fleas and even though they said they didn't have any pets and had only moved into the property recently, fleas are capable of laying dormant in carpets for incredibly long periods of time before being disturbed by the new residents.

When I visited them and took a look at their bites they didn't look anything like flea bites to me. Far too many for starters and they looked more like blisters than bites. When I asked them if they'd been to see their doctor they informed me that they hadn't – typical.

I asked if I could take a look around to see if I could find anything that might give a clue as to what the problem was. When I finally went into the bedroom I couldn't believe my eyes. The wallpaper pattern (which was quite random) was actually moving, and I wasn't even on anything. The wall was a mass of bed bugs.

Bed bugs are small, 4 – 5mm long x 1.5 – 3mm wide, parasitic insects that feed exclusively on blood and the common bed bug – which these were – prefers human blood. Their name, as you might have guessed, derives from the fact that they like to live in warm houses – especially near to or preferably inside beds. They are usually active at night, but not exclusively, and feed on their hosts without even being noticed – sneaky little devils.

Although there are a number of adverse health effects from bed bug bites including skin rashes, psychological effects, and allergic reactions, they are not known to transmit pathogens i.e. anything that causes disease, but you still wouldn't want to encounter them.

On further investigation they told me they had bought a second hand mattress when they moved in which was obviously heavily infested. I just couldn't believe that they hadn't put two and two together and realised it was the wandering wallpaper that was the cause of their plight.

The only way we could eradicate the infestation was for them to vacate the property for a few days while it was fumigated but not before the mattress was destroyed.

Chapter Twenty-One

I had never ever set an alarm and I had never ever been late. So why on that day, of all days, did I oversleep?

I glanced momentarily at the clock through sleep laden, squinting eyes and thought it read 07.30.

Surely that can't be right, (1 don't know what Shirley was doing there).

I quickly wiped the sleep from my eyes and took another, much longer, look. It was indeed seven thirty and my train was at eight thirty.

Fuckety, fuckety, fuck, fuck. I was going to be late. And as you already know, I don't do late.

I got washed and dressed – in a fashion, but not one that you'd recognise – and then drove, like a maniac, to the railway station and had to stop at every red light (traffic light, not brothel) on the way. Once there I parked the car – well abandoned it really – and then had to wait ages in a queue that was a mile long (bit of an exaggeration) at the only ticket counter open. Eventually, and against the odds, I managed to scramble, breathless, onto my train as it was pulling out of the station. Hallelujah, I wasn't late - assuming it was on time!

And I thought letting the train take the strain would reduce the stress of the journey. How wrong could I be?

I did, however, in spite of the frenetic start to the day, board that intercity train brimming with hope – and a dash of optimism – because I knew it was going to be a fantastic day. I could feel it in my water. Or was that just the cystitis?

After wandering through the busy carriages for a few minutes, I eventually found a vacant seat next to a badly scratched and grimy window. My intention was to sit back, relax and watch rural England pass me by as I hurtled through its beautiful countryside. I placed my battered old attaché case on the, somewhat sticky – it would have beat UHU (and yoo-hoo to you too) hands down – table in front of me and extracted my

Walkman from it. (I reckon it won't be long before one is featured on the antiques road show). I inserted a music cassette, of what I can't remember, and then sat back to enjoy a comfortable and stress free journey to Borntown.

The best laid plans of mice and men!

The train, it has to be said, had seen better days. In fact its better days must have been thousands of days ago. Apart from the dirty windows and sticky table – it was going to be like parting Velcro to get the case off of it – another issue was the seat fabric that had worn to a glossy shine and my seat had two large buttock imprints deeply embedded into it before I'd even sat down.

At the next station several more passengers embarked hurriedly including a very smartly dressed gentleman who asked if the seat opposite me was taken. When I informed him that it wasn't he sat himself down with an almighty sigh – or an Omnipotent Simon as he is also known.

He looked like a textbook business man – whatever one of those is. He wore a navy blue pinstripe jacket, and I assumed the trousers matched as well, but I couldn't see under the table and I certainly wasn't going to go rummaging under it to find out – it wasn't that important. His, whiter than white, shirt collar – he must have used Persil – looked is if it had been over starched because it was as 'stiff as a board' and in the middle of it proudly sat a, not very discreet, paisley bow tie. He wore wiry half rimmed glasses that were perched precariously on the end of his long aquiline nose as he read the Financial Times; well it was a pink paper at any rate, which he'd retrieved from his – not in the slightest bit battered – attaché case. He also had a terrible, but entertaining never the less, habit of talking to himself. I couldn't always work out exactly what he was saying, but I did hear the odd; "Bloody typical," and the occasional, "What the fuck." Then out of the blue – and quite loudly – "Shite and fucking derision." I don't think he actually had Tourette's, but he was only a motion away.

Behind him two young children, a boy and a girl, knelt on their seats and pulled themselves up so they could see over the headrests. They waved cheekily to me before their mother told them to sit down properly and stop annoying people. I don't think they were annoying anyone really, and certainly not me, I just hoped they hadn't heard Mr Business-Man's last remark. I think their mother might have had difficulty explaining it away.

When the refreshments trolley finally trundled through the carriage Mr Business-Man ordered a black coffee and a sandwich, which I have to say resembled the type comedians have made jokes about for time immemorial. It was actually curling up at the edges – like Aladdin's slippers. I think stale is the word I'm looking for. It also smelt like I imagined Aladdin's slippers would too – cheesy. I just hoped that it was, in fact, a cheese sandwich.

I ordered two coffees – greedy I know, but I was gasping for a drink – which was something that I would live to regret – and by the time I'd finished them both I was buzzing like a bluebottle on speed with two legs stuck in an electrical socket.

After the train had pootled along for about an hour or so, it approached the outskirts of Oxdale. As we approached the station I vaguely heard a garbled announcement over the public address system. It sounded like the announcer was in the toilet and chewing gum at the same time as speaking.

I didn't really pay too much attention to it as I thought it was only informing passengers what the next station was.

Wrong.

The train was terminating at Oxdale.

Fuckety, fuckety, fuck, fuck. And an assortment of other, purely selfish I admit, thoughts clouded my now confused and panic stricken mind.

Once disembarked on the Oxdale station platform, myself and my travelling companions (sounds like a band) were told to get on the train in front which would take us to Readbridge.

Readbridge!
But that's still only half way to my destination.
Then what?

We would be told what to do next when we got to Readbridge.

I was in the mood to tell **them** what to do – there and then.

The train in front was one of those old slam door commuter trains with only five carriages and at least four doors per carriage. A British Rail Class 308 to be precise for all you train spotter types out there.

It was crowded.

Actually it was **over**crowded and I was starting to feel somewhat hot and bothered and ever so slightly stressed. It was, after all, the hottest day of the year so far and I was on my way – supposedly – to an interview that could potentially change my life, and for the first time in my life I was going to be late.

When we eventually pulled into Readbridge, after stopping at every bloomin' station on the way, we were told to get on the next train to arrive at that platform which would then take us on to Solentown.

Another BR Class 308 train pulled up, albeit shorter than the last one, and we all unceremoniously piled on. This time we were packed in like sardines, or pilchards if you prefer, and so I spent a very unpleasant hour squashed up against someone's sweat stained, fishy smelling, armpit who hadn't discovered the benefits of deodorant.

Mmmmmmm – nice.

I've since given up that pastime you'll be pleased to know.

When we arrived at Solentown we had to change trains, yet again, this time for Borntown.

Hurrah.

When I'd looked at the map of Borntown, before I'd set off, the Town Hall looked fairly close to the train station and so I made the executive decision to walk to it. They were in fact 1.3 miles apart – and still are funnily enough.

With the combination of an exceptionally hot day, the fear of arriving late, being way beyond nervous and, after looking at my watch, starting to run, I arrived at the offices of the Environmental Health department just in time for my interview, but soaking wet. My hair was matted to my head and my clothes were stuck embarrassingly to me. I couldn't have looked more like a drip if someone had thrown a bucket of water over me. And to top it all, I was now desperate for the toilet. I knew those coffees would come back to haunt me.

"Hello, I've got an interview at 2.00 p.m., is there a toilet I can use please," I informed and then implored the frizzy auburn haired receptionist whilst clenching my pelvic floor muscles trying to prevent an accident.

How to impress people in one easy lesson – not!

At least I would be remembered, even if not for all the right reasons.

After emptying my, fit to burst, bladder – I pissed like a racehorse on diuretics – I tried to spruce myself up a little, but you know what they say; 'you can't polish a turd'.

Within nanoseconds of my arrival back at reception, feeling relieved if not relaxed, I was escorted up to the third floor and Mr Kelsall's, the Deputy Chief EHO, office.

I just hoped that he hadn't been waiting for me.

'Why are you late?'
'I needed a wee.'
'Is that why you're wet?'
'No, I ran here.'
'What, all the way from Northford?'

It wouldn't have exactly been the greatest start to an interview in the history of the world.

I was surprised, nay shocked, when I entered his office to see only Mr Kelsall languishing casually behind an unnecessarily 'large imposing' desk that took up at least a sixth of the room. A desk that only had a telephone, a computer monitor, which wasn't even switched on, and a keyboard – computer, not Casio – on it, and nothing else, diddly squat, zip.

Was this the legendary 'paperless office' I'd heard rumour of?

On either side of his desk were floor to ceiling book shelves that towered imposingly over him like Manhattan sky scrapers and were cram packed full of – well books funnily enough. All the shelves were bowing, quite noticeably, under the tremendous weight of the books giving the distinct impression that they were about to collapse. Fortunately they didn't, not while I was there anyway.

On one side were the legal tombs *Butterworth's Stones Justices' Manuals*, all arranged in chronological order, whilst on the other was a complete set of *Municipal Year Books*, again all in date order. (I think there was a bit of OCD going on there). It all made him look rather small and insignificant, even if he wasn't – well in his eyes anyway.

Mr Kelsall had a well lived in face. It was dominated by glacier gouged crevices across his forehead; deeply furrowed crow's feet – or laughter lines if you prefer – fanning out from the corners of his large staring eyes; and bags under them that could have been packed for a month's trip around India – for two.

Not sure if Ryanair would've allowed them as cabin baggage though. And yes I know, Ryanair don't fly to India – yet!

He'd seen some seriously 'good times', that was painfully obvious.

He was also losing his hair – hopefully not as a result of my arrival – and dyed what little he had left as the natural grey, making its reappearance around the temples, gave testament to.

175

I'm sure he must have shaved that morning, but a five o' clock shadow was now in evidence that gave him the appearance of a dastardly cartoon villain.

He didn't wear a jacket, but instead wore a beige – or other equally non-descript colour – cable knit cardigan underneath which I could see a badly creased checked shirt. More creased than you'd expect from just a mornings wear anyway. It didn't look as though he'd actually slept in it, but it was as wrinkled as an old man's scrotum, that was for sure. (I'm an old man now so I know these things).

The office window had been left slightly ajar allowing a refreshing breeze to waft over me, cooling down my perspiration – because only horses sweat – and drying me out, but not in the rehab sense of the word.

So there isn't going to be a panel of 'experts' to intimidate and interrogate me then? How very strange.

But that wasn't the only strange thing about the situation. The way Mr Kelsall was sat in his chair was a bit odd to say the very least.

His left leg was draped casually over one arm of the chair and his knees were bent and spread wide apart while his feet were positioned close together.

Having your 'legs akimbo' I believe is the correct terminology.

And no, before you ask, it wasn't a yoga pose. A pose it was, but not a yoga one I can assure you!

The chubby fingers of his calloused hands were laced together at the back of his head as he leaned back casually in his high backed chair, and to top it all, his trousers were alarmingly tight as well. I couldn't avoid staring at his 'lunchbox' and could see that meat and two veg were most definitely on the menu.

I don't know if it was nerves or the breeze from the window, but all of a sudden my throat constricted and my lips went incredibly dry. But the last thing I was about to do was lick them and then swallow, whether it improved my job prospects or not.

Pleasantries were exchanged as I was asked to take a seat and make myself comfortable and then the interview began.

Is Mr Kelsall going to rearrange his legs into a more appropriate position so that I don't have to stare at his crotch?

No, it would appear not.

He gave a quick scratch of his 'lunchbox' – which gave a whole new meaning to the term 'finger buffet' – and then asked his first question.

"Why did you want to be an environmental health officer?"

That was the first question I was asked when I went for the student job. Is it really relevant now?

This time, however, I just went for it and stated the most obvious answer, even if it wasn't the one he wanted to hear; 'I wanted to save Wales', I confessed.

Only joking!

"Why do you want to work in Borntown?"

I must have come up with a suitably impressive answer to that and the other questions I was asked, because at the end of the interview I was offered the job.

Now that the more conventional questions were out of the way he went on to ask some less conventional ones including, "What makes you angry?"

Stupid questions like that...

I later learnt that if you didn't answer the question he would ask what you'd do if you saw someone kicking a dog.

If it was a Rottweiler, I'd call an ambulance.

I think he thought he was some sort of psychologist.

He wasn't.

After many more 'unconventional' questions he finally asked, "Now do you have any questions?"

Naturally I assumed, and we all know what that does, that he was talking to me. Little did I know that William May, who would be my line manager at Borntown, was sat in a corner behind me – completely out of view – and piped up, "What Aids training have you had?"

Well it frightened the living daylights out of me. Not the actual question, but unexpectedly hearing another voice in the room, and a big booming one at that.

William May was a muse for Mr Kelsall, or he provided him with amusement would be more accurate. Whenever William wanted to do something new and innovative he would go, with me in tow, and ask for

Mr Kelsall's permission, who would inevitably say, "No." At that point William would throw a childlike tantrum, stamping his feet, whinging and whining – 'the whole nine yards' (to use the colloquial American phrase) – after which Mr Kelsall would say, "Yes." How futile.

Sorry, I've digressed, back to the interview.

Then another strange thing happened. I know, how many more could there possibly be?

Mr Kelsall's secretary waltzed in – I later learnt that her foxtrot was rubbish – and asked who wanted a cup of tea.

Ten minutes later we were sat having tea and biscuits as though we were long-lost buddies. Unfortunately there were only rich tea biscuits – there wasn't a chocolate digestive in sight. It was probably for the best though as the chocolate would have invariably melted all over my fingers and where would I have wiped them? I didn't dunk the biscuits either as rich tea are notoriously pathetic for dunking. You need a more robust biscuit, like a ginger nut, for that.

As I've already mentioned – 374 words earlier – I was offered the job there and then and accepted it without hesitation. I would have time to make my mind up properly whilst I was waiting for the contract. I must admit that I was justifiably suspicious that no one else had applied for it, but was informed, when I started work there, that that was not the case.

Initially I was appointed on one spinal column point down from the top of the pay scale, which was a considerable increase for me and was more than welcome to cover the increased cost of living on the south coast.

A week after my interview Mr Kelsall telephoned me to say that unfortunately my lease car was going to cost me more at Borntown than it did at Northford.

How can that be? They are just taking the lease over.

As a result he told me that he would be putting me on the top of the scale to cover the additional cost.

What a nice man.

It turned out that he wasn't that either.

I thought that he was strange at the interview, and in the words of Bachman Turner Overdrive: 'You Ain't Seen Nothing Yet' – and I certainly hadn't.

He was the original – and best – control freak. He would check everyone's car mileage claim each month and if he didn't believe one, or more, he would drive the route(s) to see what the actual mileage was, dispute half a mile – at 6p a mile I hasten to add – and then claim the mileage himself.

Was that a cost effective use of his – very expensive – time? I think not.

He would also randomly check on telephone calls made to see if anyone had been using the office phone for personal reasons. Once he requested the pleasure of my company – although not by formal invitation – because someone in my team had phoned America and he wanted to know who and why.

After minimal investigation I discovered it was a free phone number that had been called to a training company – in America as he'd quite rightly pointed out – to book him on a course that he wanted to attend.

Put that in your pipe and smoke it.

He was also vindictive.

I'd been off sick for a couple of days and when I returned to work I asked for a flexi day off the following week. He refused the request point blank. When I asked him why he told me I could take the day off as annual leave but not as a flexi day even though I'd accrued enough time to do so.

See what I mean?

More than a little disgruntled I went to see the Chief about the matter who told me that I was more than capable of sorting the problem out for myself.

While I was there, and in the mood – a bloody foul one – I asked the Chief why it took so long for Mr Kelsall to let people enter his room when they'd knocked even though no one else was in with him. He told me – in all seriousness – that it was to give him time to put his Game Boy away in the drawer. And by 'Game Boy' I mean the electronic toy.

Now my gander was well and truly up. I stormed back to Mr Kelsall's room, knocked on the door again and waited for the usual amount of time before he shouted, "Enter."

Before I was even half way through the door I accused him – quite viciously – of refusing my flexi day request as a punishment for having taken sick leave.

He looked totally aghast at my brazen cheek and reversed his decision immediately and let me have the flexi day after all.

What a complete and utter waste of time. A grown man (physically if not mentally) playing power games – on his 'Game Boy' as well probably.

Just goes to show you though, if you stand up to a bully more often than not they back down.

Chapter Twenty-Two

One of the reasons I left Northford Council was for promotion. I thought it was about time I started to climb the dizzy heights of the career ladder – with someone holding onto the bottom of it obviously. The other reason was to specialise, and so I was going to be a Senior EHO at Borntown Council specialising in education and training.

Little did I know at the time that title means very little, if anything, and it certainly doesn't make anyone perform the job any better. I've lost count – because I only have eight fingers, two thumbs and ten toes – of the number of meetings and courses I've attended where, at the start, the group were customarily asked to introduce themselves and say where they were from. Invariably there was always some over inflated ego – that you could have attached a wicker basket to and sold balloon rides in – who would spout forth a job title that was as long as your arm – after being stretched on the rack – that would never have fit on a standard business card in a month of Sundays. They would then, boastfully, proceed to tell everyone what they were responsible for, which would take at least another five minutes – if you were lucky, longer if you weren't – that usually just fell short of negotiating world peace.

Not in the least bit necessary.

Just do as you were asked.

You're only that important in your own head.

For goodness sake, get a life.

It reminds me of the poem *The Indispensable Man* which suggests that if you fill a bucket with water and put your hands in it and then pull them out again, the hole that's left behind is how much you'll be missed. Especially in the work place! The last sentence was mine, not the poets.

I don't know of anyone whose dying words were, 'I wish I'd been more important'.

We are all indispensible.

We are all replaceable.

Get over it.

Another reason why I moved on was for more money - moula, wonga, dosh – whatever you want to call it. Although the increase in salary was soon swallowed up by the cost of living on the south coast, most notably the price of houses and therefore the mortgage payments. But the main reason was because my new job entailed education and training, the two aspects in environmental health I enjoyed the most.

How lucky was I?

Very few authorities had officers focusing solely on those areas, and I think even less do now, and anyway, you know what they say, 'Those who can, do; those who can't, teach.' So I was going to teach. (Unfortunately I've come across far too many incidences over the years that have proved that point admirably).

Bitch!

But it's true, Salchester College being a prime example.

I've also heard of an addition to that quote which is: 'Those who can't teach, teach teachers.'

Don't shoot the messenger!

Alternatively William Arthur Ward said: 'The mediocre teacher tells. The good teacher explains. The superior teacher demonstrates. The great teacher inspires.'

I've had proof of that saying as well during my numerous years in education and at the risk of blowing my own trumpet – even though I don't actually own one – I was regularly told that I explained things really well when I was teaching. So maybe I was a good teacher? I'd like to think so.

On my first day at Borntown I was thrust into the thick of it, and my new health education role, because William May – he of the big booming voice in the corner behind me at my interview – was launching the department's 'Heartbeat Award'. This was not an award given to people who regularly watched the 1960s based police drama series set in Yorkshire, but a scheme set up by the Health Education Authority (HEA) and the CIEH to encourage eateries, nationally, to offer healthy options on their menus. It wasn't just aimed at hotels, restaurants and cafes, but take-aways and workplace canteens as well.

The criteria to gain the 'original' award were: (i) the premises complied with the food hygiene regulations. (At one stage it was muted that this requirement was to be removed. So you would be able to purchase a healthy meal in a filthy, vermin infested, premises. I think not.)

(ii) 33% of the options on the menu were healthy; (iii) 33% of the staff were trained in food hygiene and; (iv) 33% of the dining area was designated a no smoking area. Needless to say – but I'll say it anyway – the fourth requirement has been superseded by the current smoke free legislation.

The launch was held at a very posh four star hotel in the centre of Borntown that quite a few 'celebrities' stayed at when they were appearing in the town. The hotel had 'donated' the room we were using for the launch free of charge and included a sumptuous buffet lunch – if not a particularly healthy one – as well.

Decide for yourself: it included pate and a selection of salads doused in fatty dressings as well as gooey cream laden desserts guaranteed to clog the arteries and increase the heart rate. I think the five a day message must have passed them by because there was no sign of a vegetable crudités or a piece of fresh fruit. Not even a grape.

Whoever was tasked with writing out the labels for the food made an almighty faux pas as well as they described a mousse as a mouse.

Unless of course it really was a mouse!

At least they didn't put Moose.

We suspected that the hotel thought because they'd hosted the event they would automatically receive the award. Well they had to think again. When the community dietician, Yvonne – more about her in Chapter 27 – looked at their regular daily menus, they didn't meet requirement (ii) – they weren't even close. Their chef, a bit of a prima donna to say the least, argued that when people ate out in hotel restaurants, like theirs, they wanted rich and extravagant food that they wouldn't normally cook at home, food that was cooked in butter and had cream and alcohol added to it. We couldn't really dispute that, but equally there were some people who wanted an alternative. William gave the example of business men who regularly ate business lunches and had been told by their doctors that they needed to cut down on the rich food and loose a few pounds, as well as a few inches. All we wanted the hotel to do was offer some healthy options. For example if chips were on the menu, then offer boiled potatoes as well. Was that really too much to ask? Apparently it was.

After several months of silence, (it was a battle of Wills – assuming the chef was called Will as well), we did eventually hear from him again. He said that he had made some changes to their menus and they were now offering a selection of salads and other low fat dishes which were going down a treat with their business clientele.

Now, just because it says salad doesn't mean to say it's actually healthy. The dressing alone could contain a disproportionate amount of fat, which would deem it unhealthy, alone – as with those at the launch. And how low was their low fat anyway? Would it manage to get under the bottom rung on a limo pole? Not that I was sceptical, but for the award to really mean anything at all and be worth the paper it was printed on – which was pretty good quality, 300gms – the criteria had to be applied rigorously.

When Yvonne looked at their new menus the changes they'd made were as radical as they were dramatic and easily enough for them to have achieved the award.

I didn't gloat.

Over time some proprietors criticised the 'original' scheme saying that if they exceeded the requirements i.e. 50% of the menu was healthy or 100% of the staff were trained in food hygiene, there was no extra recognition for their efforts. Which I have to say was a very valid point. So as a result, when I got the grant from the Health Authority – see Chapter 27, again – I decided to launch our very own version of the scheme to address the issue.

Borntown's scheme – which had the very catchy title; 'The Borntown Heartbeat Award' – had three levels: Gold, Silver and Bronze. The Bronze Award was basically the 'original' scheme with a requirement of 33% in each category. N.B. the smoke free legislation had not yet come into force. For the Silver Award premises needed more than 40% of the food to be a healthy option, more than 70% of the food handlers had to be food hygiene trained and more than 70% of the dining area had to be non-smoking. The Gold Award increased those values to more than 50% healthy options, 100% food handlers food hygiene trained and 100% no smoking area.

What amazed me was that proprietors only ever wanted the Gold Award. They were not willing to compromise and put up with second best, and most definitely not third, and would do whatever was needed to get the Gold Award.

By introducing our own scheme I believed that we would:

- increase consumer choice
- increase the healthy choices in as many outlets as possible

- grow the scheme so that the logo became universally (not literally) recognised throughout the Borntown area – some hotels actually used the logo in their advertising
- enable people to eat in a 'clean' environment, i.e. complying with food hygiene regulations

The award was valid for 12 months, after which re-assessment was necessary – a bit like an MOT really. Spot inspections could be carried out at any time and if any of the criteria were not being adhered to, then the award would be rescinded.

It was left up to the district officers to promote the scheme when they were carrying out their routine inspections, especially to those premises that already complied with the food hygiene regulations, but this turned out to be a bone of contention as most of the officers were reluctant to promote the scheme because they felt it wasn't part of their remit. They just couldn't see health in its broader, more holistic, sense and it was incredibly difficult to get the blinkers off them. I think it would have needed an anaesthetic.

The saying, 'You can't teach an old dogs and new tricks' came to mind.

We even evaluated the benefits of the scheme at the local council run leisure centre. The catering manager there was initially concerned about the perceived loss of income from making any changes to their menu i.e. not offering the traditional, high selling, items such as cheese and chips. Also, how the menu changes would be welcomed, or not as the case may be, by the customers. Her eventual approach was to make small changes to the cooking methods and increase the healthy choices gradually, without drawing too much attention to them. In the end there was an increased profit because the healthy options were so popular and because the cheese bill was cut by 70%.

A regular – yearly – health education campaign that I got involved with was national No Smoking Day, and if my memory serves me correctly – it's not gone self-service yet – it was initiated by the HEA. Or was it called the Health Education Council back then? Forgive me if I'm wrong. I put it down to stupidity.

Each year there would be a different theme associated with the day and a different tag line and logo, although the overall message was still to

'Stop Smoking'. Here are just a few examples of past campaign slogans: 1994; 'Kick the Habit', 1996; 'Put a NOT in it', 1999; 'Take the Plunge'.

In 1990, for some unknown reason, there wasn't a theme or logo produced. I don't know why. Perhaps it was due to a lack of funding. So my colleague and creative genius James Lax came up with the 'Smoke busters' campaign, which was an obvious play on the *Ghostbusters* film's characters. He designed the outfits – yes we dressed up – which consisted of a white paper boiler suit, a utility belt complete with ray gun and plastic safety goggles. Well it made a change from the French maid and nurse's outfits!

If you want to have a bash at making the outfit for yourself, all the components are easily available from your local DIY store.

We, there were three of us, also had a plastic jerry can strapped to our backs with a funnel for the 'quitters' to put their cigarettes – not a dimp ('something valuable' like money or 'a stupid person') or a butt ('a backside' – can you imagine getting yours into a jerry can?) – into. For each cigarette 'donated' they would receive a raffle ticket and the chance to win one of our many – I'd like to say spectacular, but they weren't – prizes. Like the hype?

Now you're dying to know what the 'unspectacular' prizes were, aren't you? Come on now, be honest. There's no shame in it. Well I'd tell you if I could remember, but I can't. However, they must have been vaguely desirable because we had a lot of interest and a lot of cigarettes to dispose of at the end of the day. They probably included free passes to various health and fitness centres in the area. They would have certainly been something 'healthy'.

James also produced our very own comic book featuring the adventures of the 'Smokebusters' in a battle with their arch enemy Nico Tine – as opposed to his brother Oval – and his addictive nature.

I wonder is there an organisation or support group that helps people who are addicted to the night time malted milk drink?

The comic also contained startling facts about the effects of smoking, handy tips on how to give up and information on where to get help quitting. Unfortunately I was killed on page 4. I wasn't quite sure if that was and subliminal message and actual wishful thinking or just the way the cookie crumbled. Some questions are just better left unasked.

In 1994, as I've already mentioned, the message was 'Kick the Habit' and the slogan certainly lent itself to a campaign involving football. By this time I'd given the No Smoking Day event to the student EHO to organise as health education and promotion was one of the elements in their logbook.

The student that year, a cheeky little chappy who always had a mischievous glint in his eye and went by the name of Lesley Lancaster – I was sure that he'd gone by some other, altogether unrepeatable, names as well during his twenty years on the planet, but I didn't want to know what they were – contacted the local football club, then in division 1, to ask if they would like to be involved. Fortunately the answer came back as a resounding yes. The manager even gave a radio interview about the benefits of not smoking in sport and the team posed for photographs to be printed in the local press on the actual day.

Lesley, or Les as he preferred to be known, was a complete and utter joy to work with. What he lacked in stature he certainly gained in personality. He always had a sunny disposition, always saw the positive side to every situation and managed to elicit an element of fun in everything he did. You might say he was the perfect pick me up when life had put you down. He was certainly a breath of fresh air in an environment that had a lot of stagnant air circulating in it from some of the old fuddy-duddies.

No Smoking Day was always held on a Wednesday – in March. I don't know why and never took the time to find out either. But as you can imagine, a Wednesday was not the best day of the week for such an event because a lot of people were at work. Add to that the fact that the month of March could be wet – in fact most months in the UK could be wet – and cold, not many people were out and about to receive the message.

Because of the problem with Wednesdays, Les had an absolutely brilliant idea. "Let's do our No Smoking Day event on a Saturday instead," he suggested.

"Go on, tell me more," I encouraged.

His genius of an idea was to hold our event at the local sports centre on the Saturday morning prior to No Smoking Day, at the football academy where there would be loads of people.

The majority of the crowd would inevitably be children, and mostly boys at that, but we thought that the message 'not to start smoking' was just as important, if not more so, as stopping. I always thought, in my naivety, that if we concentrated on stopping people starting smoking in the first place it would reduce the need for No Smoking Day altogether. Prevention, after all, is always better than cure. Or is that just too obvious?

As part of his master plan, Les created a game of skittles where the skittles were made out of the inner tubes from rolls of carpet and made to look like cigarettes. All the children had to do was kick a ball at them to knock them over; which was easier said than done I can tell you. Also the goal keeper from the local club came down and the children had the opportunity to take penalties against him – which they absolutely loved doing, and a few of them were considerably better than some of the England players!

In 1999 the slogan was 'Take the Plunge' and for a publicity shot the student, Victoria, asked me if I would jump off the top diving board in the local swimming pool. She did ask me, politely, she didn't tell me to go and do it, I hadn't upset her – I don't think. She also wanted me to dress up as a cigarette – no not a fag. Then after she'd asked me, she decided to ask if I could actually swim! Nice afterthought. There was no mention of whether I was scared of heights.

How could I refuse?

To coin the anti-drug slogan I could; 'Just say no.' But I didn't.

Some years the slogans were just far too obtuse to do anything with. It was always claimed that they'd consumer tested them on the public, but I have to admit that, with some of them, I thought the public must have been extracting the Michael if they said that they actually liked them.

Another of the 1980s/90s new campaigns that I got roped into – sorry involved with – was 'Drinkwise'. This was ostensibly about highlighting responsible drinking and promoting the recommended number of units of alcohol a man and a woman should drink per week – now a daily amount – to avoid future health problems.

Trying to get those messages across without being accusatory and implying people had a drinking problem or leaving them feeling bamboozled was going to be a challenge of epic proportions – well not that easy anyway.

The first event I organised was simple enough in its concept and was held in the town centre. I managed to acquire, after very little persuasion, a selection of low alcohol and alcohol-free drinks from a variety of local outlets more than willing to promote their wares. These included beers and wines as well as the more obvious soft drinks that were made into a selection of mocktails – alcohol free cocktails. The idea was to encourage people to substitute their regular alcoholic beverage for the low or alcohol free alternative from time to time – we were certainly not promoting abstinence. Also, it was to let the public have a taste of these, relatively new on the market, products so that they could discover for themselves that the alternatives tasted just as good as the alcoholic versions. Except that wasn't always the case. In fact some of them were disgusting, but there have been improvements made over the years.

I could argue that the campaign was an unmitigated success, but I am not that naive. Certainly a lot of people showed interest and we did get rid of all the free samples, but it was a hot summer's day and I'm sure that most of the punters were just after a free drink to quench their thirst instead of buying one from one of the cafes. It would have been interesting to see if there was a drop in cold drink sales that day. I just hoped that a few people picked up on the 'health' message as well.

The following year I attempted something completely different – a pub quiz.

'What's so different about that?'

I'm glad you asked me that question, because I was just about to tell you.

Firstly all the questions were, in one form or another, loosely related to drink and not necessarily alcoholic ones either, but liquid refreshments in general. For example:

Round 1: Name the famous TV pubs. We gave the teams the names of various popular TV programmes such as; Emmerdale Farm, Eastenders and Coronation Street (the more obvious and relatively easy ones) as well as Only Fools and Horses and The Simpsons and all they had to do was name the pub or bar that was featured in each of the programmes.

Round 2: Guess the drink from the advert. We used current, full page, adverts taken out of popular weekly/monthly magazines and they had to name the drink that was being advertised. The Gordon's Gin advert at the time consisted of just a green page.

Now I know what you're thinking, how did I know it was green? Easy, if it's a solid block of colour, like a green field or full page ad, I can see it. Alternatively, someone – who wasn't green/red colour blind – could have told me.

We even used an advert for milk – yes we were that dastardly – which no one got.

Round 3: Name the drink from the bottle's description i.e. 'a complex blend of some of Jerez's finest wines: delicate finos, aged amontillados, and fragrant olorosos, with wine from the special Pedro Ximenez grape providing a silky, mellow smoothness.' There was one point for the actual name of the drink and one point for the specific brand.

We also had two taster rounds where each team had to nominate one member to represent them and taste, funnily enough, a selection of drinks.

Round 4: Taste the difference (a great name for an up-market range of supermarket foods, wouldn't you agree?). Each contestant was given an alcoholic wine, a low alcohol wine as well as the alcohol free version, and then the same with lagers, to taste and then distinguish which was which. Would you be surprised to learn that no one got them all correct?

Round 5: Guess the alcohol content. Again each team nominated a representative to taste a selection of liquids, including milk – which really confused some people – and then place each one in a band of alcohol content i.e. 0 – 1%, 1 –3% 3 – 4% etc. One of the liquids was the original Woodward's Gripe Water which I was staggered to learn had an alcohol content of 3.6%. Guinness was only slightly stronger at 4.1%. You will be relieved to know, however, that the alcohol has since been removed from the gripe water. No wonder babies went off to sleep after a dose or two.

'What'll you have to drink?'

'I'll have gripe water on the rocks please.'

Round 6: Units of alcohol etc. This was one of the more serious rounds where we asked the teams' questions about the recommended number of units of alcohol per week for a man and a woman, the legal blood alcohol level for driving in the UK and other such sensible questions. Every team got all these questions correct. So that meant we'd either done a fantastic job at getting the message across or they were just clever and knew it already. I'm ruling out cheating and opting for the former – obviously.

There were other rounds, ten in total, but I won't bore you with the details – assuming you're not already bored. I'm sure you've got the gist – for which I believe you can get ointment.

On each table we provided the teams with a bottle of red, white and rosé low-alcohol wines courtesy of Marks and Spencer and the bar was serving free alcohol free lager courtesy of Badger Brewery. Other soft drinks were available too.

My 'Smokebuster' colleague James was the quizmaster – more Robert Robinson that Jeremy Paxman – whilst I frantically marked the answer sheets between rounds.

It was a terrific laugh, and I wished all health education events could have been like that. All the teams said that they'd had a fun as well as an informative evening and hadn't minded in the slightest that it was alcohol free(ish).

But I bet they all went to the pub after it though!

One of the most constant, frequent, regular, and annoying – and totally unnecessary – complaints the department received was about dog fouling. Or to put it in everyday parlance, dog shit.

'Pooper Scooper shit is gonna find me,

But I won't feel blue,

Like I used to do,

Cause somewhere in my pocket there's you,' – sung to the tune of Abba's Super Trooper.

Because the subject was such a high profile and political 'hot potato' I was asked, well told – if truth be told – to develop a campaign to address this thorny issue.

Obviously you can't really liken doggy doos to a hot potato, apart from them both being hot when they come out of the 'oven', and as far as I'm aware it doesn't usually contain thorns either, but I like the metaphors.

So in an effort to get some inspiration, and pass part of the buck, I contacted the local art college and asked them if they would like to be involved in designing the materials for our 'Responsible Dog Ownership' campaign, as it was eventually called – after agonisingly long deliberation. Fortunately they enthusiastically agreed and said that they would use the opportunity as a design brief for their second year students.

I looked at many designs – all of them good – but eventually chose the artwork that had the cutest dog and the simplest slogan: 'For everyone's sake, pick up the mess he makes.' The message was printed on posters and stickers of the dog sat on the pavement, the beach and in a children's playground.

We also ran a competition in the local newspaper to name the dog and the winner was SYD – Supervise Your Dog.

The college also produced a short video for us called 'Pavement.' It was filmed looking down on a garden gate and showed the garden path and pavement on either side of it. It was a time lapse film showing a number of people walking through the gate and up the garden path as well as along the pavement. Towards the end of the ninety second film a dog comes along and performs on the pavement in front of the gate – I don't know if it had an equity card – followed by the postman who then treads in it and then looks up to the camera. It was so clever and summed the issue up in such an effective way.

One of our more unusual and rather unique campaigns in the early 1980s was focused on the, in my opinion inflammatory labelled, 'Aids epidemic' – as the press were so fond of describing it. What we wanted to do was create a greater awareness of the disease and how easy it was to avoid contracting it, which would scotch some the countless, and quite ludicrous, myths surrounding it and therefore hopefully allay people's misguided fears. This was at the same time the Government had their television and postal 'Don't Die of Ignorance' campaign.

You remember the TV advert: the image of an erupting volcano followed by a tombstone being etched with the word AIDS, all designed to – and in a lot of cases succeeding in – scaring the public half to death.

Our campaign had a far less obtuse message and consisted of a set of three A5, hand drawn, stickers aimed at particular dates – we thought the figs were safe enough – in the year. The stickers were distributed to all the pubs and clubs in the town to be displayed in their toilets, both male and female – which they did. Result.

The first one was of a Christmas scene with a decorated tree in the foreground and a collection of clothes, male and female, strewn across the floor leading to an open bedroom door. The strap line was: 'Do you know what you're getting this Christmas?'

I guess we should have done two different ones on the same theme really. The other one having two sets of male clothing, but we didn't. I think it would have been a step too far for the council at that time, being far too risqué.

The next showed a list of New Year's resolutions which included; 'Always carry a condom', emphasising that it was the responsibility of both males of females.

The third and final one in the set declared; 'On 14th February V D stands for Valentine's Day. On 15th February it might stand for Venereal Disease.'

The local rag, allegedly, contacted Barbara Cartland (she was alive then, it wasn't done through a medium – or even a large for that matter) for a comment on our sticker. She is supposed to have said, and I paraphrase, that she was disgusted that we could turn such a romantic day into something so smutty. That was the point Barbara dear, romance can lead to other things and without the necessary protection it can lead to something else entirely. In the words of Ed Sheeran: 'Love can hurt.'

I wanted to do one for Halloween as well which said something like: 'If you're going to put the willies up someone – wear a condom.' Which I thought covered all angles – if not positions. But I was voted down on that one – never really understood why.

We always tried to evaluate the success of the campaigns we ran, but it was always very difficult.

It was easy on a quantitative scale; we could say they were successful because a lot of people turned up, we spoke to a lot of people and a lot of information was distributed. It was relatively easy on a qualitative scale too because we could get people to fill out evaluation forms. But you could never measure the effectiveness of the message, i.e. how many people actually changed their behaviour as a result of what they'd learnt. In reality only time would have the answer to that question. That didn't mean that we should stop trying to influence though. Even if one person changed a habit – and they weren't a dirty monk – then it had to be worthwhile.

Chapter Twenty-Three

"You are not going to believe what I've just brought upstairs," announced the boss's secretary, in a sort of stage whisper, as she burst into the room and then dramatically spread her arms across the door. More pantomime than Pinter it has to be said. She'd been to collect the first candidate of the day to be interviewed for the student environmental health officer post.

"What do you mean," I responded apprehensively, wondering what the hell was waiting for us outside the door.

"Just you wait and see," was her all-knowing reply and with that she smirked, asked if we were ready and then escorted him in.

The 'him' in question was Lewis, a young man who, according to his CV, had gone straight to university from school to study environmental health.

What can I tell you about Lewis?

The person who shuffled, and I do mean literally shuffled like a decrepit old man, into the room looked like a mere child. I kid you not – pun intended. He looked as though he shouldn't have been let out on his own. The clothes he was wearing must have belonged to his father or some other male relative because they surely couldn't have been his. They were at least two, maybe even three, sizes too big for him. I know when I was bought a new school uniform it was always a touch on the large side to start off with because I would 'grow into it' as my dear mum used to say. But he was twenty, how much more growing was he likely to do? The sleeves of his jacket were almost completely covering his hands and his trousers were crumpled up around his ankles. His shirt collar hung around his neck and gave the impression that his head had been shrunk. He reminded me of Tom Hanks at the end of the film *Big* when he gets his wish to be a boy again. (Well I know it wasn't Tom Hanks at that point but a young David Moscow, but you get the picture).

I offered my hand in welcome and shook what can only be described as a pathetically limp and disgustingly clammy appendage devoid of any vitality.

Yuk!

Following the obligatory introductions and customary offer of a seat I got the ball rolling by asking him how his journey had been.

"Okay," I strained to hear.

"Did you travel here by car?" I continued to probe.

"Yes."

His second reply was no louder than the first.

"Did you drive?"

"No."

"Did you find us easily?"

"Yes."

I could see I wasn't going to be able to get a word in edgeways!

I always liked to start an interview by asking the candidate to tell me a little bit about themselves. An easy enough question to relax them and get them talking before I asked the more serious and relevant questions.

"So, tell me a bit about yourself."

"Like what?"

Two words now.

"Anything you want."

"My name is Lewis and I'm studying Environmental Health at university."

A full sentence!

Really, tell me something I didn't know.

Silence.

Is he going to say anything else?

The silence continued unabated.

Obviously not.

"Okay then, let's move on."

I glanced at my fellow interviewer who was smiling away inanely, but wouldn't make eye contact with me. Probably for the best really as we both would have ended up laughing at him, because there was no way we could have laughed with him. No matter how many open questions I asked, he somehow always managed to reply with a yes or no answer. How was he going to cope when he was a qualified officer dealing with the public, landlords, business owners? I wasn't quite sure that environmental health was really the career for him and never found out why he'd chosen it when I asked him the question either.

Bearing in mind I'd allocated half an hour for each interview, we'd finished his in just fifteen, tortuous, minutes, give or take a pregnant – with the gestation period of an elephant – pause or three. When I eventually asked him if he had any questions for us, he looked completely shocked at the very suggestion and then proceeded to ponder for several minutes. Or had he just gone into a trance? It was very hard to tell. Would we actually get another complete sentence out of him? The suspense was killing me. In the end we were treated to his inevitable reply, "No."

Oh well, at least I tried, which is more than I could say for him.

Actually, I was now tired of him – a nice anagram there.

The third candidate – I won't bore you with the details of all six of them, just the ones that stood out, and not necessarily for the right reasons – was a young woman by the name of June Bramshaw, who was dressed somewhat informally for an interview. By that I mean she wore a short, flimsy, floral patterned, wrap over skirt, a skin tight vest, without a bra underneath – June was bursting out all over – and flip flops. She looked as though she'd just walked in straight off the beach.

Not sure that the floral bit is relevant, but I've told you anyway. At least you've got the complete picture. Nudge nudge, wink wink, say no more.

She informed me proudly that she'd arrived in Borntown on the Monday – it was now Wednesday – so she could "get a feel for the place", as she so eloquently put it. Good idea I thought, she's actually doing some research. But I soon realised that I was being unrealistically optimistic – silly me.

"So why would you like to have your placement here with us?"

"You've got a beach."

Silence.

"Is there anything else?" Thinking she may refer to the public health challenges of a seaside town and those peculiar to Borntown.

"No, not really."

I thought I'd try another approach.

"What can you tell me about Borntown?"

"You've got a beach."

Silence.

Ever get that feeling of déjà vu?

So she had just come straight off the beach after all.

I then enquired if she had any more interviews lined up to see how proactive she was being at getting a placement.

"I did have one at Salchester tomorrow," she sneered, "But I've cancelled that one because I want to come here."

That was the wrong move, because you certainly won't be coming here – that's for definite.

Last, but certainly not least – size wise at least – was Chris, who actually got the job. He wasn't the greatest of candidates, but because the job hadn't been advertised until the end of May for an August start, and I know this sounds mean – but it's true never the less – we didn't exactly get the pick of the crop to choose from. In fact there were quite a few bruised specimens and some, I think, had actually gone off, but he was easily the best of the bunch. And I'm not sure if that's a compliment or not. At least he'd done some research and knew a bit about us as an authority and the public health challenges we faced.

Chris reminded me of Billy Bunter, certainly in size if not in character. He was fat. There is no other way of putting it. In fact he was so fat that his cheeks (bum) hung over the sides of the chair he was sitting on. He also had a strange habit of straightening pictures on the walls of the houses he visited. He was told, in no uncertain terms, to stop it because it was an invasion of privacy. I thought it was a bit creepy too.

Interviews are daunting enough when you are being interviewed, but believe me they are not a piece of cake – Battenberg or otherwise – when you are the interviewer instead of the interviewee.

When I became the training officer at Borntown Council, after William May left, the department only sponsored one student throughout the length of their three year degree course, providing practical training during their breaks from university. That was way back before the advent of the logbook. As you're already aware, because you've been paying attention, environmental health is now a four year course.

When the logbook was launched (God bless her all who sail in her) I decided, unilaterally I admit, that because students only really needed a placement for their practical year out to complete it, we would just have a student for that year. This meant that we would be able to accommodate a student every year instead of every three as we had been doing.

Fortunately I'd already been on a two day recruitment and selection training course, so knew what I was supposed to do – a rare occurrence – to effectively recruit and select staff.

I always held the interviews on a Wednesday. Not absolutely sure why now, but there must have been a valid reason originally, I expect. Perhaps it was because it was in the middle of the week or that it was the only day of the week with a W in it. Your guess is as good as mine.

The interview panel consisted of me (I know, bet you didn't see that coming) and the administration and finance manager. I wanted someone with me who hadn't got an environmental health background and was therefore totally independent to prevent me from being unnecessarily biased. The other reason was in case I got asked any finance and administrative type questions at the end of the interview.

Not as dumb as I look!

The interviews were always held in the department's library. Well I say library, but I use the term in its loosest possible sense. The room was so small you couldn't swing a cat around, (a 'cat o' nine tails and not a real one obviously) and only had a couple of shelves with books on. The books consisted mainly of a selection of very old, well ancient actually, and outdated environmental health tomes. They were covered in a thick layer of dust, which only went to prove how (un)frequently they were referred to, and could only have been kept for their historical value. Or should that read hysterical.

I always liked to arrange the seats in a triangle, with everyone facing inwards (very cosy) and without the ubiquitous 'large imposing' desk between them and us, for the interviews I was hosting. Fortunately our knees didn't quite touch in the overly cramped room. I personally believed the arrangement negated the feeling of hierarchy because there were no barriers between 'us and them' and it therefore created a much more relaxed atmosphere. Well that was my intention anyway.

As I was preparing the room the first time round it suddenly dawned on me how weird it was to suddenly have the shoe on the other foot. Obviously I quickly changed it over because it was really uncomfortable and I'd heard that was how you got bunions. In all seriousness though, it was now me doing the interviewing and not being interviewed – for a change.

It felt totally bonkers and just a teensy weensy bit scary!

The following year I was much better organised and sent out the job advertisements in the November. The closing date was the last Friday of January and the interview date the second Wednesday in February, both of the following year – obviously.

That year we were inundated with applications, the majority of which were of a very high standard. This made it all the more difficult to choose between them. So in the end I had no option but to shortlist eight instead of the usual six.

I find it strange what people are willing to tell you on their applications, and why they think it's relevant to the job they're applying for. For example: 'I make a good cup of tea.' Where in the job description does it mention making tea? Anyway, what constitutes a 'good cup of tea' is a matter of opinion. I have friends who like builder's tea, the type that you can stand your spoon up in. I also have friends who just waft the tea bag over the receptacle containing the boiling water and as a result have tea that has barely got the strength to get out of the mug – or cup if you're posh. But they all have one thing in common: they all believe that theirs is a 'good cup of tea.'

'I was the local country dancing champion three years on the run.'

What are you trying to tell me? That you have other interests or you have a competitive nature? Or perhaps even both.

The day before these interviews the candidate who was scheduled to be on first, at 09.30 – and lived locally – rang up to ask if she could rearrange her interview for another day because she went to university on a Wednesday.

"No," was the short answer.

Well you can't make it any shorter than that really.

"You knew the date of the interview from the advert that was sent out last November," I emphasised the November and was surprisingly polite even though I was actually seething. "The interview will be over by 10.00 and then you can go off to university. I'm sure they'll understand."

Not exactly the way to make friends and influence people. The candidate that is, not me. I gave up trying to do both a long time ago.

Another student came armed with a clipboard – would probably be a tablet now-a-days – to which was attached a sheet of fluorescent yellow A4 paper on which I could see a rather long list of questions.

Fair enough though, at least he was well prepared.

I started the interview the same way I always did, but was immediately interrupted by him asking me if he could ask me some questions.

"There'll be plenty of time at the end of the interview for your questions," I explained, but he still kept interjecting throughout the whole interview.

"Is it my turn now?"

"No not yet."

I was beginning to wonder exactly who was interviewing who.

I couldn't believe that some of the students hadn't the slightest idea about interview etiquette. Why did they think that their rudeness would impress the very people who were going to make the decision on whether to appoint them or not. You may find it hard to believe, but he was one of those who were not appointed. I know, I'm just too darn picky. And his questions, when he did eventually ask them, were trivial and irrelevant.

A year later and I was going through the whole process yet again. I couldn't believe how quickly time flew when you were having fun. And believe me, for the most part, I did have fun at work.

I loved being involved with student training. It kept me on my toes most of the time – which is why I have such large calf muscles now – and I had to keep up to date with current information and practices, not only in case the students asked me questions, but because I was a conscientious EHO as well – naturally. I can distinctly remember one instance, however, and there were many more obviously, when one student asked me a question that I didn't know the answer to.

Alan was writing a report for his logbook after a visit to the local landfill site. He wanted to specifically refer to the leachate, the liquid that runs off the dump, and asked me what I knew about it.

I was totally honest with him, "Absolutely nothing," I declared, not in the least bit embarrassed.

Well the look he gave me was priceless. You know that 'you're supposed to know everything' look we gave our parents when we asked them questions they couldn't answer. Or should that be wouldn't answer?

But most of the time I did answer their questions. To the best of my knowledge anyway – and you can take that either way.

Sorry, I've digressed; let's get back to the interviews.

Do you remember the woman, a few pages ago, who lived locally and wanted to change her interview the previous year? Well she was shortlisted again. I know, you must be disappointed that she didn't get the job last time, but that's the way the cookie crumbles I'm afraid – especially an oat and almond one.

Anyway, can you guess what she did this time round?

Correct.

She asked exactly the same question about moving her interview, to which she got exactly the same answer. I honestly thought that I was having yet another, but quite different, Groundhog Day.

This time though, during her interview and the ensuing discussion, I said something to which she venomously objected, "You didn't say that last year."

We are talking about a mature student here who had a family, not some immature kid straight out of school. I was totally nonplussed.

And no she didn't get the job – again.

I used to tell the candidates that I would inform the successful one the following morning by 10.00. Therefore, if they hadn't heard from me by then, they unfortunately hadn't been successful.

Usually it was obvious who'd got the job – by gut instinct mainly but then confirmed by using the scoring system. There were never any of those 'too close to call' situations where we would have had to go to a tie-break – not that I had a system in place for that anyway, so I would usually call the 'winner' later the same evening.

Some of the reactions I got to the news were unbelievable. One female screamed down the phone unintelligibly, with pleasure I hasten to add and not in agony, for a good few minutes. She was in total disbelief that she'd actually got the job. Another one was so laid back about it that he sounded as though he was completely disinterested. I later found out that that was his natural reaction to most things – good or bad.

If any of the candidates wanted to contact me afterwards for feedback I was always willing to provide it and quite a few of them did, but they didn't always necessarily agree with my comments. I was told, on more than one occasion, that I had made completely the wrong choice. Well, on the basis that they hadn't met the competition and didn't know who I'd appointed, I didn't know how they could say that.

Chapter Twenty-Four

To fully qualify as an EHO and be registered with the Environmental Health Officers Registration Board (EHORB – no longer The Environmental Health Officers Association as it was in my day) – I think it's now changed to the Environmental Health Registration Board (EHRB), I can't keep up, but what's an 'O' between friends – not only did you have to have a degree in environmental health, but you were also supposed to pass all three of the assessments set by the CIEH.

I say 'supposed' because there are plenty of EHOs out there, employed by scores of local authorities, who have not yet completed/passed all of these assessments. Many chief officers don't even know what it takes to become an EHO anymore and frankly my dears I don't think they give a damn.

Okay, I'll get off my high horse now – if someone will fetch me the steps.

These assessments consisted of: the logbook, the interview and a batch of professional, written, exams. I know, it was a bit long-winded and I never really understood why students had to sit more written papers when they'd already achieved their degree. The professional exams, however, have now been replaced by a 'substantial written case study' which concentrates on environmental health interventions. They must have read my mind – which wouldn't have taken very long.

Now, because you've been exceptionally good and got this far in the book – unless you're just skimming through it of course – I'm going to give you a well-deserved treat and tell you about some of my escapades as a logbook and interview assessor.

Once upon a time, long, long ago...

When I was a student, along with Edwin Chadwick (Google him), and had completed my practical training, the local authority were expected to sign a formal declaration to say that their student had experienced all aspects of environmental health and then submit it to the CIEH.

Unfortunately there was no verification system and inevitably some authorities told porkies.

Surely not!

I know it's hard to believe, but it's true. Get over it.

This lack of foresight was eventually put to the test when a newly qualified EHO was employed in the housing section of an environmental health department, but had never actually inspected a house in their life.

Oooops!

So because this requirement was being laboured an alternative delivery method was sought and after a short gestation period, the Logbook was eventually born. It was a painful birth, especially without the aid of pethidine, and there were a few teething problems as well – yes, it was born with teeth – but it was so desperately wanted.

The Logbook was basically a National Vocational Qualification (NVQ) portfolio. It stated the number and type of visits and inspections that had to be carried out and how to write them up for assessment.

Over the years it's gone through more than one metamorphosis. It's changed its colour from the original bright blue to dark green – so I believe – and then to a deep burgundy. (NB none of these colours are featured in the Farrow and Ball colour chart – the names are far too obvious). I think that was the correct sequence anyway. The requirements have also been changed and updated with each change of colour.

It's even changed its name along the way too from the Logbook to an Experiential Learning Portfolio or ELP and now the Portfolio of Professional Practice or PPP.

Now to me, ELP stands for Emerson Lake and Palmer, who were my favourite band back in the 1970s. It's now Elbow – obviously. Which reminds me, did I ever tell you about the time I got on stage with ELP and shook Keith Emerson's hand? No? Well I was fifteen years old, quite lanky, thin as a rake and very spotty – they actually used to call me 'pizza face' – and I'd gone to see them with a group of mates at the local concert venue. Half way through the gig, give or take an epic fifteen minute song or two, everyone rushed to the front of the stage, so at that point I thought I'd try my luck and see if I could climb up on to it. As I was desperately trying to pull myself up I suddenly felt a pair of cupped hands underneath my foot giving me a peg up. I assumed it was one of my mates, but when I looked either side of me I could see that they were all in view, so it was obviously someone I didn't know. How considerate was that? When I eventually stood up on the stage, a larger than life – and meaner looking than treble the meanest thing you could ever imagine – bouncer ran headlong towards me. I thought I was going to be unceremoniously flung off the stage in the style used in the Rolling Stones film *Gimme Shelter*. But no, Emerson must have said something to him because he suddenly turned round and strode purposely off the stage, so I managed to shake hands with my first musical hero. I then stood triumphantly at the front of the stage, with my arms in the air, and gave the V sign – no, not 'The Longbow man Salute' – to a crowd who cheered and screamed in total ecstasy at my good fortune and then jumped off. Magic!

Now where was I?

Unfortunately I've reached that age when my train of thought often leaves the station without me.

Oh yes.

I became an assessor mainly so that I could advise Borntown's students exactly what the CIEH expected from them and not what my interpretation of their guidelines was. The CIEH, however, insisted that if they'd invested the money, time and effort to train you, then the least you could do in return was work, as an assessor, for them from time to time. Fair comment I suppose. And that's how I became a logbook assessor – simples.

Not only did the students submit their – in some cases interminable volumes of – written evidence to be assessed, but some of them would also send objects that were intended to prove that they'd actually been on a particular visit. Some of these were hilarious. One student even sent a piece of coal with their logbook which obviously meant they'd visited a coal fired power plant – because you can't get coal anywhere else! Someone else sent in a faecal specimen pot as proof they'd investigated a case of food poisoning. Fortunately it was empty. So had they really done the investigation?

Each section in the logbook told the students exactly what they needed to do i.e. you must inspect three of a particular type of premise, and yet some students would only inspect two and then wonder why they hadn't passed. D'oh!

There were many students who had obviously worked incredibly hard and produced excellent logbooks, including most at Borntown I'm pleased to say. Others just did the bare minimum to pass. But at least they did do it I suppose, because as I've already mentioned some practising EHOs still haven't – and probably never will.

For my sins – and there must have been a lot of them if this was my punishment – I also trained as an interview assessor as well; for exactly the same reasons as becoming a logbook assessor.

Whereas the logbook training was done on the job, as it were i.e. sitting with a qualified assessor during a scheduled marking session and 'getting the hang of it', the interview training was much more formal held over a weekend in a hotel in Shropshire. I'd just moved from there!

On the Saturday morning we had the customary talk on how to conduct the interview and how to mark the student's performance. In the afternoon we had a series of mock interviews where we would all take turns at being interviewers and interviewees. On the Sunday though, we had real students to practice on who also gained valuable experience of the interview process.

The actual interviews were held at different times throughout the year at a variety of different venues up and down the country. This was to accommodate the students so that they didn't have to travel too far – even though some of them still did.

The interview panel consisted of two fully qualified EHOs. The CIEH always put the panels together so there was inevitably a skills and experience variance between the two EHOs.

The interview process was as follows:

On arrival the students would have their bags and mobile phones confiscated so that they couldn't take any notes in with them and there could be no contact with the outside world – so they couldn't cheat. They stopped just short of insisting that they sat the exam naked.

On entering the exam room the student would be introduced to their assessors, asked to take a seat and then the interview process explained to them. They should have already known what it was, but it was always best to reiterate – just in case they'd only recently returned to earth from another Galaxy, Milky Way or other cosmic named chocolate bar. They were then handed an A4 manila envelope containing three scenarios – or complaints if you like.

Once they had opened the envelope and checked that they had the same scenarios as the assessors – which was always useful in an exam situation – they were then left in the room all alone, completely by themselves and with no one else there, for twenty minutes. Not nineteen and a half or twenty and a quarter even, but twenty on the dot. During that time they were expected to study the complaints and then prioritise them; i.e. say which they would deal with first, second and third and justify the reasons for their choices. The students, for some unknown reason, always thought that there was a definitive answer to the priorities, but there never was. If the student came up with a plausible and strong enough argument for their decisions then that was okay with the assessors. I have to say that some of the students were very creative indeed and came up with solutions that I would never have thought of in a month of Sundays – or a fortnight even – but were valid nonetheless. Clever sods.

It just goes to show that the phrase, 'you learn something new every day', and the proverb, 'you're never too old to learn' are absolutely true. However, the proverb, 'you can't teach an old dog new tricks' is not, because you can – I'm the living proof.

After they'd explained their priorities and answered any subsequent questions like, "What would you do if your third priority was a councillor complaint?"

Now if the public health significance of the councillor's complaint is less than their number one priority complaint then the answer is to 'tell the councillor to take a running jump'. Well explain why you can't deal with their complaint first. However, in reality most councillor complaints get dealt with first just because it's a councillor complaint.

There's no need to tell me that that's wrong, I know.

After the priorities bit they then had to choose one of the scenarios to talk about in greater depth, but it didn't necessarily have to be their number one.

This part of the interview tested them on their knowledge of and understanding of environmental health processes i.e. how they would deal with the complaint having consideration of the relevant legislation and guidance as well as any council enforcement policies. It also tested their decision making abilities by demonstrating what options were available to them to deal with the situation and which one they would choose given the facts of the matter, and why.

The latter part of the interview always consisted of questions about wider environmental health issues like Agenda 21 (never found out what happened to the previous 20 agendas) sustainability, and other hot topics of the day.

As you would expect some very strange things happened during the interviews and so, for your delectation, I will give you just a selection of some of my more memorable experiences:

The students were always collected from reception by one of the assessors and then escorted to the examination room. On this particular occasion it was my turn to do the honours.

On entering the reception area I was faced with a right motley crew who were all waiting expectantly whilst looking a tad on the nervous side – to put it mildly. When I called out the name of the student I was there to collect a rather forlorn looking young woman, with pre-Raphaelite red hair and freckly skin, stood up and walked dejectedly towards me. Hoping to cheer her up I proceeded to exuded all my natural charm – the latest designer fragrance – and jovially launched into my well-rehearsed spiel.

"Hello, it's very nice to meet you."

"Thanks."

"How are you?"

"Okay."

"Welcome to the CIEH. Have you been here before?"

"Yes."

Did you have a good journey here?"

"Yes."

"Have you travelled far?"

"No."

"Isn't it a beautiful day?" And it was.

I'm British – discussing the weather is obligatory.

By the way, what's the weather like where you are today?

"Yes."

I think she might have been a bit distracted!

Not one to be easily deterred I continued, "When we get to the examination room I'll explain the interview process to you."

"Oh there's no need to do that," she said having finally found her tongue – between her cheeks, facial that is – "This is my fourth time here."

Ooooh deeeaaar.
Is she really so bad that she's failed three times already?

I wasn't looking forward to this one little bit.

When we reached the examination room I introduced my fellow assessor and we continued our friendly chat. We discovered that she'd been a technical officer working in the housing section of her environmental health department for ten years before she'd decided to become an EHO. In the previous three interviews none of the scenarios had been on housing.

When she finally opened the envelope, and looked at the three scenarios, her face immediately lit up like a Christmas tree – the one in Trafalgar Square. If you haven't already guessed, there was a housing one in there at long last, and unsurprisingly that was the one she chose to talk about in greater depth. When we started to question her on it, it was apparent that what she didn't know about housing you could have written on the back of a postage stamp – and a Bolivar 10c green at that!

Obviously she passed with flying colours – or straight A's as we called it.

Then there was the time I interviewed the prettiest of students who had the most beautiful porcelain complexion and the daintiest of features. She did look like a doll – but not in the American sense of the word. She was painfully shy and very quietly spoken – in fact a whispering mouse would have been louder – that when she started to speak I could hardly hear a word she was saying. She looked as though she was lip-synching the words to a backing track that someone had forgotten to switch on. She must have realised my predicament – cupping my ear like an old folk singer was probably the giveaway – because she quite softly said, "If you can't hear me, I'll speak up."

Again, what she didn't know wasn't worth talking about. My only concern was that if she had to give someone a bollocking (an environmental health technical term) for not complying with the law, would they even realise that that's what had happened. I very much doubted it.

There were some, however, that were just plain hopeless. No point in dressing it up. They had done little, if any, preparation and as Benjamin Franklin once said: "By failing to prepare, you are preparing to fail." How true, and I saw the evidence of those wise words far too often. One such example was a student who, when we moved on to wider context questions, was asked about healthy eating and specifically the work of a well-known – au naturale – TV chef to get children eating healthy school meals. After I'd asked the question, they just sat motionless for a moment or two and then started to systematically interlace their fingers on their lap while staring blankly into the middle distance. After a minute or two more of complete silence they started to wring their hands distractedly in their lap. I guessed at that point that they had no idea who I was referring to or what I was talking about. When I eventually mentioned the name Jamie Oliver their giggling response was, "Oh yeah, I really like him."

Was that an answer to the question?

Interviewing five students a day for between 30 and 45 minutes each, over a three day period, was really very tiring. Much, much more than I'd ever anticipated before I'd started doing them. And no, I'm not looking for sympathy either because I know I won't get any.

We, clearly, had to listen intently to every word the students said to us because what they told us was what we would to have to base our decision on as to whether they passed or not.

But let's face it, if we're brutally honest, most of the time we may hear, but don't actually listen to, what people say to us. Am I right, or am I right?

To keep ourselves entertained during some of the more boring interviews, some of the assessors would play silly games – including me. A bit childish and not very professional I know, but sometimes needs must. One such game was to give each other an obscure word that we had to incorporate into one of the questions.

I can remember quite clearly giving my colleague the word Womble. That had to be impossible to use surely. Well it would appear not. When it came to the wider context questions he asked the student about waste and recycling. "Can you tell me what you understand by the Womble principle?" he asked with a perfectly straight face. The unsuspecting student, however, didn't even flinch at the term. They understood exactly what was meant by the 'dubious sounding' principle and started to talk, at length, about reducing waste; if possible re-using it; and if not, recycling it.

Absolutely brilliant!

We would also, on the odd occasion, do other things to keep ourselves amused, especially if we'd had a succession of poor students to contend with. On returning to the examination room from reception with the next student, my 'Womble' colleague introduced me as his 'life partner'. I have to admit that it was a struggle not to laugh. The student, however, was so uptight about their situation that it went straight over their information-crammed head.

Although we were paid by the CIEH – £15.00 per logbook and £15.00 per interview – the money wasn't the motivation; it was actually nice to be able to give something back to the profession. I do realise that you've probably got your fingers down the back of your throat after that comment, but I mean that – most sincerely folks.

The proof of that, as if proof were needed, is that a bad logbook could easily take a day and a half to mark because feedback had to be given on everything that was wrong with it. Alternatively a good one could take as little as three hours.

The CIEH did look after their assessors though. We, more often than not, stayed in good hotels and in the evenings would be taken out for a nice meal to some really lovely restaurants. It was always good to wind down after a long day with a glass or two, and if we were lucky, with something in them as well.

It was also good to meet up with old friends and colleagues that you'd assessed with before, but also to make some new acquaintances too who's knowledge you could tap into when you had problems back at your own authority that was in their field of expertise. What was known as networking I believe, to use the appropriate jargon.

One of the hotels we frequently stayed in, however, did give us regular cause for amusement.

Most of the assessors would arrive on the Monday evening, before the assessments started on the Tuesday, and have dinner at the hotel. It was much easier to do that on the first night because people were arriving at different times. However, if our meal happened to drift on beyond 21.00 the chef would go home and we invariably didn't get a pudding. Not even strawberries and cream – and they were on the menu – which you would have thought the waiting staff could have rustled up. But no!

It reminded me very much of the *Fawlty Towers* episode with the Waldorf Salad.

At breakfast on one particular morning at the aforementioned hotel, I asked the waiter if I could have poached eggs on toast – I really fancied them for a change. And a change is as good as a rest.

"Sorry, we don't do that," I was told quite officially.

Not wanting to cause a scene, but still wanting the eggs on toast I enquired, "So do you do poached eggs?"

"Yes."

"And do you do toast?"

"Yes."

"So can I have poached eggs AND toast?"

"Certainly," and off he went into the kitchen.

Ten minutes later he actually served me poached eggs on one plate and the toast on another.

The look on his face when I actually put the eggs ON the toast was priceless. I am such a rebel, even if it is without a cause.

The assessments, both logbook and interview, were also regularly held in Northern Ireland to save their students from travelling, and I always loved going over there to work. The only problem was that a lot of their legislation was different to that in England, Scotland and Wales, but there were always local EHOs on hand if we had any queries.

On rare occasions their students would offer quite convincing reasons why they hadn't managed do some of the inspections they were supposed to. However, when we asked the local EHOs if they were legitimate reasons we were always reliably informed that they were not reasons at all, just feeble excuses and that we should fail them for not doing what was asked of them – which we did.

On one of my many visits over there, after a particularly tiring days work, we all decided that we should go on a jolly and visit the Giants Causeway. So we piled into the two cars owned by the two local EHOs and they drove us the fifty or so miles to the famous World Heritage Site.

It's a good twenty minute walk from the car park to the actual causeway, but well worth the trek. The 40,000 interlocking basalt columns – I counted every single one – are not only stunningly beautiful to witness, but also a fascinating geological phenomenon to behold. Mother Nature certainly knows how to impress, that's for sure, and the view, over to Scotland – just 13 miles away – was absolutely breathtaking. Well, it was for the first ten minutes we were there anyway. As we stood admiring it, and our magnificent surroundings, it suddenly started to get obscured by a rapidly moving bank of dense clouds that headed in our general direction. Within five minutes we were completely engulfed by them as they started to systematically deposit their aqueous contents all over us – it pissed down. By the time we got back to the car park we were completely drenched – having not taken any coats with us because the weather was so nice when we first arrived.

We were booked into the local hotel for dinner that night and when we walked in, dripping water everywhere, the maître d' just smiled knowingly at us and said, quite matter-of-factly, "Been to see the causeway and got caught out by the rain have ya's?" I think they must have been used to it.

As we sat eating our delicious dinner, the steam started to rise from us as we started to dry out. I bet we all smelt like wet dogs as well, although no one actually complained.

One of the classic moments though was with a student who I did a mock interview with. Their performance was more than a bit of a worry to say the least.

After discussing their prioritisation of the three complaints – and passing that element – the scenario they chose to talk about in greater depth was concerning a little old lady who was very timid and very nervous of authority. She was, however, having regular visits from the district nurse for an ongoing medical condition and had built up a rapport with her and trusted her implicitly. The nurse was concerned that there might be a problem with dampness in the privately rented property and contacted the local environmental health department for advice on the matter.

I asked the student what they would do first.

"I'd have a look at the past history of the property," was their confident and correct answer.

"Anything else?" I questioned.

"Ask other officers in the department if they know anything about the property."

It was all going swimmingly.

"You are informed that no-one has ever been to the property before and there isn't an existing file on it either. Now what are you going to do?"

"I will have to go to the property to see what the problem is for myself."

"Okay, so talk me through what you would actually do."

"I'd go to the property and ask the old lady if I could see the dampness."

"And how would you get in?"

"I'd knock on the door."

"And say what?"

"I'm from environmental health and I want to see your dampness."

"Okay. So what if she says she won't let you in?"

"I'd tell her she's got to."

"And has she?"

"Yes."

Not actually true. 24 hours' notice has to be given if it's a private residence.

"What if she still won't?"

"I'd go and get a warrant."

I asked the student how they'd go about getting a warrant, but they didn't know.

"Okay, let's move on then," I encouraged. "Let's assume you've got the warrant. What would you do next?"

"I'd go back to the property and tell her she has to let me in now because I've got a warrant."

"And what if she still won't, what would you do then?"

"I'd kick the door down."

My jaw just dropped. I couldn't believe my ears.

At that point the interview was stopped. There was nothing else they could say or do to rectify the situation. They had failed, and with at least a seven figure minus score.

I am sure, dear reader that you have worked out for yourself that they should have gained access to the property with the district nurse whom the old lady trusted. They could have been introduced as a colleague and then made the appropriate assessment. Can you imagine the headlines in the national papers: 'Old lady dies of heart attack as council official kicks door down'.

Chapter Twenty-Five

The worst part of the job, without a shadow of a doubt, was being on call.

This right royal pain in the proverbial involved being available, out of hours, for a whole week starting on a Friday evening at 17.00 and finishing the following Friday morning at 08.00.

This service was meant to cover 'environmental health emergencies' only, but because the term was never clearly defined, we inevitably went out to almost anything.

The urgent shrill of the mobile phone cut through the still night air like a hot knife through butter. My bleary, out of focus, eyes could just about make out that the time on the clock radio read 00.30, or was it 00.38?

A knot tied itself in the pit of my stomach and tightened as I grabbed the phone from my bedside table and pressed the connect button.

I was informed, by a far too bubbly voice for that time of night/morning – never quite sure which it is – that a car alarm was sounding incessantly outside a hotel on the outskirts of Borntown. Not only was it keeping all the guests in the hotel awake, but the residents of the area were disturbed too. And that wasn't the first time I'd heard that either.

I never slept particularly well when I was on call. I usually just drifted in and out of consciousness throughout the night and always felt like sh1t the next day. I was always worried that I wouldn't hear the phone go off and, therefore, miss a crucial call. Even so, it was always a shock when it did actually ring – like all calls are in the middle of the night.

I made a note of the details of the complaint and then fell out of bed – literally. I stumbled in the dark, like a paralytic drunk, to the bathroom to do an adrenaline-induced nervous poo before eventually fumbling to put my clothes on – surprisingly the right side out and the right way round.

In no time at all I was heading towards town on what was a balmy – in fact it was completely crazy – summer's night. There wasn't a cloud to be seen in the sky and I was totally mesmerised by the multitude of stars that twinkled effortlessly in the vast ebony expanse. They looked just like the sequins, hand sewn of course, that constantly glittered on one of those sparkly Latin American dance costumes.

When I finally arrived at the hotel there was an almighty kerfuffle going on outside it. It looked as though the majority of the guests were milling around the car park all clad in their not-so-sexy nightwear. And what a diverse and interesting array there was too. The majority of it was made up of fabrics that are now rarely used, including rayon and winceyette, in a mind boggling range of gaudy colours and jazzy patterns. Some of the garments were a tad on the tight side, to say the least, and left very little to the imagination – which wasn't a particularly pretty sight I can tell you. I thought I'd stepped into the pages of a down market lingerie catalogue. You know the one, 'Phallus in Undieland'.

Snob!

After being bombarded by the guests with demands to "Do something about this bloody noise," etc. – I did something.

On investigation I established that the offending car was actually parked in the hotel car park. I duly contacted the police and gave them the registration number of the vehicle to find out the name of the owner of it and where they lived.

The guests were still loitering, but without intent, even though I'd told them all to go back to bed as there was nothing they could do – except get in the way that was. But the British, as well as having an insatiable appetite for queuing, are an inquisitive lot. Well bloody nosy really. They do love something to gossip about, or even better still – a scandal.

It transpired that the owner of the vehicle lived in South Wales, so either the car had been stolen or the owner was actually staying at the hotel. My money was on the latter, even though I'm not a betting man. I asked at reception if anyone was staying at the hotel that had the owner's name. The night porter rummaged through the register, rather lackadaisically it has to be said, and then after what seemed like forever muttered, "Yes." I then asked him if the hotel took the registration numbers of guest's cars when they checked in, even though I knew the answer to the question, and rather embarrassingly he muttered, "No."

Well that's not particularly useful now is it? But there was no point in taking it out on him, it wasn't his fault if the management were incompetent and didn't have the simplest and most obvious of systems in place. (Please note that the opinions expressed are mine and not the publisher's).

It emerged that the owner of the car had requested a quiet room at the back of the hotel, away from any noise – how ironic. So the first he knew about the problem was when the night porter rang his room and asked him to switch his car alarm off.

So not all the guests were awake then!

He promptly arrived downstairs minutes later only to be met by jeers and boos (mine's a gin and tonic please) and countless, very choice, derisory comments, as he attended to his car.

How embarrassing. I bet he left early the next day – even if he was meant to stay longer.

I eventually arrived back home at 02.30, disrobed, threw my clothes on the floor in a 'teenage' crumpled heap and then climbed back into my, now stone cold, bed. I played bed angels for the next few minutes to warm it up again and then just as I was starting to drift off, into the land of nod, the phone rang again. This time it was the police. They wanted me to attend a noisy party at a block of flats in the centre of town.

I've just got back from there, I wanted to scream, but professionalism got the better of me. And if you believe that, you'll believe anything.

So it was back to the bathroom for yet another 'you know what' and then I robed once more before setting off on another thirty minute drive back into town.

It really is quite interesting the diversity of 'sights' you see in the wee small hours of the morning, all bathed in that weirdly unnatural haze of the sodium street lamps. Not to mention some of the sounds.

Some of the sights I can tell you about, whilst others I can't – or won't to be more precise.

I'm absolutely convinced that there are more urban foxes now than their rural cousins if the number I saw that night were anything to go by. I felt really privileged, as well as fascinated, to watch these beautiful mammals roaming the streets as if they owned them, rooting through plastic bin bags for their next meal and then playing like kittens as if no one was watching.

There were also a surprisingly large number of drunks tottering back home, completely loaded after painting the town red – no comment – and all performing that 'one step forward two steps back' dance that they do. They all looked like marionettes that had had at least one of their strings cut loose.

I got to the flats not long after 03.00 and was met by the police who were wearing protection (you're making your own story up now) in the form of stab jackets, and were probably armed with their trusty truncheons too (stop it). All I had was a flimsy (not the designers name) fleece and a tatty PACE (Police and Criminal Evidence Act – in case you were wondering) notebook. The fleece wasn't even high visibility either.

I was so wound up by then that I thought I was having heart palpitations until I realised I could actually feel the music thumping on my chest. It was really belting out from the first floor flat and even with the windows closed it was loud enough to wake the dead – metaphorically speaking of course. It sounded like Duran Duran – *Planet Earth* (bop bop bop bop bop bop bop bop). I could also see numerous flashing coloured lights. No it wasn't the start of a migraine. Although I think the combination of the noise and the lights would have brought about a severe headache given time. It looked as though they'd hired some professional disco equipment.

They were having a right old shindig up there.

I knocked on the door several times before I was heard over the general hullabaloo emanating (is that impersonating Eminem?) from upstairs. The door was subsequently opened by a skinny young oik who had a thick band of white paint plastered across his nose and cheeks; a tricorn hat plonked precariously on his head – which was going to be fat in the morning – and a cape on (no, not a castrated cock – although if he didn't do as he was told...).

It was clearly an eighties fancy dress party – or at least I hoped that that was all it was.

He slumped against the door jamb for physical – and maybe even moral – support, speechless, and then stared way beyond me, at God only knows what, with his severely blood shot, eye lined, eyes. If his name had been Anthony you could have called him 'A Dumb Ant'.

Please yourself.

I think he might have been ever so slightly inebriated.

"Can I speak to the tenant please?" I enquired rather briskly.

"Ooze asin?" he slurred, still not looking at me. Or perhaps he thought he was.

Translation: Who is asking?

His breath was a pungent cocktail of alcohol and cigarettes. I was convinced that if I'd struck a match close to his mouth the fumes would have ignited and he would have given a pretty reasonable impression of a dragon.

"I am," I replied, fairly politely given the circumstances.

"An oo d'ya fink y'ar?"

T: And who do you think you are?

There's no need to get lippy!

"I don't think, I know who I am."

"Ooze at 'en."

T: Who is that then?

"Your friendly, *" but not for too much longer if you continue to piss about,* "neighbourhood environmental health officer," I said with a hint of annoyance. (Great name for a colour).

"And what's your name?" I asked, not expecting a particularly coherent answer.

He looked me straight in the eye and then started to giggle like a pre-pubescent schoolgirl.

"Bates."

One of the policemen wrote it down in his notebook.

"Of course it is," I said sardonically. "Would that be Master Bates by any chance?"

He giggled even louder.

The policeman crossed the name out.

"Sa woya wan?"

T: So what do you want?

"You're making far too much noise."

"Sa wha d'ya wan mi t'd?"

T: So what do you want me to do?

"Make less noise."

"No."

T: No.

Or words to that effect.

"Now, now sonny" interrupted the other copper, "there's no need to be abusive. Just do as the nice man tells you and we'll be on our way."

A bit presumptuous I thought. How did he know I was nice?

The youth stumbled back up the stairs, slipping back down a few of them more than once. It was like watching a human game of snakes and ladders – and he certainly wasn't winning. I then heard the music being turned down the minusculist of amounts before he fell, arse over tit, back downstairs.

He gradually picked himself up of the floor, looked around furtively to see if anyone had noticed – just like a cat would do when it falls – and then stood up again, all bolshie like, in the doorway – without his hat.

"That's nowhere near enough," I informed him thinking of the bruises he was now going to have in the morning to go with the fat head.

"Okay, shwa desbl d'ya wani?"

T: Okay, so what decibel do you want it?

"Look, I haven't got a noise metre with me and I doubt if you have one either," I retorted. I was beginning to lose my patience. "If I can still hear it in the street then it's still too loud and therefore a nuisance as far as I'm concerned."

"An if I doe d'wa ya say, wa ya gona d?"

T: And if I don't do what you say, what are you going to do?

"Seize the equipment."

I didn't tell him that I would need to get a warrant to do that. I didn't think he'd have understood what I was talking about, not in that state anyway. Also the process was long and protracted and could have quite easily taken me a couple of hours to execute. Something he didn't need to know either.

I would have had to contact the out of hour's clerk to the court to get a Justice of the Peace out of bed to hear why I needed the warrant. If they agreed with my reasons, they would then sign and issue it. I just hoped that my bluff would work. Smoke and mirrors, it was all just smoke and mirrors.

The change in his attitude was quick, but he sobered up even quicker. It was obvious that they'd hired the equipment and would be in even deeper doo-dos if they didn't return it after the weekend.

There was always an Achilles heel. All you had to do was find it.

He dashed back upstairs, two at a time this time, without tripping up or falling down once and turned the volume down to what I considered to be an acceptable level. This was followed by a loud chorus of disapproval and then after a short, but terse, exchange of words he came back downstairs.

"Is that okay mister?" he enquired – his diction having improved considerably.

"Yes," I said, "As long as it stays like that. I don't want to have to get out of bed to come back and deal with this again."

"Did you have to get out of bed to deal with this then?"

Isn't that what I just said?

"I did," I said in a suitably irritated tone. "But don't worry; I don't really like sleep." The sarcasm was starting to kick in now.

"I am so sorry mate for any inconvenience we've caused you. Really I am."

I couldn't believe the change of tune – it was now Spandau Ballet's *Gold*. I was expecting at least some verbal abuse, if not an attempt at the actual physical. That's why the police were there on this 'sting', (get it). But no, he was as nice as pie – apple pie obviously, because gooseberry isn't that nice at all.

Who'd have thought it? It was now obvious that he was wearing completely the wrong outfit because he'd actually turned out to be a proper little 'Prince Charming'.

By the time I got back home I was tired, so very, very tired, and prayed that I wouldn't get another call that night – which thankfully I didn't.

On another occasion I was called out, around midnight again, to a car alarm that was sounding in a different hotel car park, but this time they knew that the owner was staying there. Unfortunately though, the owner had gone into town for a night out and there was no way of contacting them. Unsurprisingly the hotel didn't have a mobile phone number for them.

I served a notice on the car, even though it couldn't read it, giving the owner one hour to disable the alarm. I then sat in the reception area of the hotel waiting, and waiting, and waiting for the minutes to tick away and the notice to expire. The night porter just kept giving me filthy looks as if to say, 'what are you doing, besides nothing', but I had to follow procedures. (Makes me sound like a real jobsworth, but in my defence, these were legal procedures). If they didn't return within the hour I could legally get someone else to disable the alarm or alternatively have the vehicle towed away.

With only ten minutes to spare the owner walked in through the hotel door.

"Mr Herbert?" I enquired.

"Yes."

"Can you hear that noise?"

"You can hear it half way down the street," was the cocky reply.

"Do you know what it is?" I enquired again.

"Sounds like a car alarm to me," was his second and even cockier reply.

It was hardly Mastermind.

"Can you be more specific?" I asked.

"I don't know what you mean."

"I'll give you a clue," I said helpfully and then quoted his car registration number.

"That's my car," he said.

Give the man a coconut.

"So it must be your alarm too," I informed him.

Red faced, as if he'd been under a sun lamp for too long, he bolted into the car park and with the last few seconds of the notice expiring, switched it off.

The last car alarm story is even shorter.

When I got to the car the alarm had actually stopped sounding. I could see that someone had forced the bonnet open and I assumed that they'd cut the wires or more likely wrenched them out. At least the problem was solved, even if it wasn't legally.

We also dealt with house alarms.

Again we had to serve a notice on the owner/occupier by attaching it securely to the property if no-one was in. The notice would be placed in a waterproof sleeve to protect it and usually state that the noise had to be abated immediately. Alternatively a specific time period could be given, which was determined by the officer, i.e. within twenty minutes.

More often than not the owner/occupier was away on holiday and so the notice was never going to be complied with.

On rare occasions the complainants would say that the alarm had been sounding for days. So why did they leave it so long to complain and why wait until the middle of the night? I always wanted to ask.

Alarms should only sound for twenty minutes before they cut out and the alarm should be serviced on a regular basis so that this happens. The owner/occupier should also nominate a key holder and inform the police who it is so that the alarm can be switched off, without this palaver, in similar circumstances. Unfortunately that happened very infrequently.

The next step was to get a warrant to enter the premises, a locksmith to open the door and make the property secure afterwards and an alarm company to disable the alarm as well as a police officer to witness that it was all being done above board. What a faff. Trying to coordinate all that was like trying to spin plates in one of those 'Cirque Du Soleil' circus acts. So I thought I would take an easier option, just get the alarm company to cut the wires to the outside box.

After serving the notice I went to the office and starting thumbing through the yellow pages to find an alarm company that would attend in the middle of the night – because at the time we didn't have a contract with a specific company. Needless to say it was much easier said than done. Some had no out of hour's numbers – useful. Some when they eventually answered the phone said that they wouldn't attend at that time of night, but would be there first thing in the morning – *you're awake now so you might as well come now* – whilst others wanted to know how they would get paid. It was all very helpful.

After I don't know how long, but I'd grown a beard, I managed to get someone to attend who would accept an invoice payment – hallelujah. By this time the residents must have thought I'd abandoned them.

The alarm was eventually silenced from the outside and the really nice bloke informed me that I was lucky because if there'd been an internal alarm as well, that would have still been sounding.

The owners, when they eventually returned home a week later, were none too pleased to find a letter through their letterbox informing them that their house alarm had been disabled while they were away and with a bill attached for the pleasure.

"You had no right to do that," was their vexed assertion when they rang me to complain about the complaint.

"I have every right," I informed them, "and consider yourself lucky that it didn't cost more." I then went on to explain how I had saved them the cost of a locksmith as well. That shut them up.

Councillors were a pain in the neck when it came to the out of hour's service. Correction, councillors were a pain in the neck. Okay, not all of them, but most of them. It was on a Saturday evening in June, at about 19.00, when one rang me to say that his neighbours were hosting a 21st birthday party in their back garden – a back garden that backed onto his back garden.

And?

Thanks for letting me know, but I think I would have survived without that information thank you – just saying.

Of course I didn't say any of the above.

"It's going to go on until 23.00," the councillor protested.

"How do you know that?" Was he psychic?

"It says so on the invitation."

"You've been invited?"

"Yes."

"But you haven't gone?"

"No."

"So what's the problem?"

"It's noisy."

I couldn't believe I was having the conversation.

"I haven't received any other complaints."

With that the line went dead.

It was obviously a one off event. And as I'd told him, I hadn't received any complaints from any of the other neighbours – probably because they were all at the party – so it couldn't have really been that much of a problem.

He called again at 23.00 on the dot. "It hasn't finished."

"Well I'm not coming out. The out of hour's service is for emergencies only and this is not an emergency. It's a one-off, special occasion, birthday party and I still haven't had any other complaints from anyone else in the area."

I was convinced that he would put in a complaint about my lack of action and my attitude so on the Monday morning I told the boss exactly what had happened.

We heard nothing.

He probably realised what a proper party pooper he was and hopefully what a total prat (or a word that rhymes with it – yes you're right, brat) he was.

Another of our 'out of hours' duties was the Friday and Saturday night noise patrols between the hours of 22.00 and 02.00. This was to monitor the regular offenders. During one of my Saturday night stints, I was called to a noisy party.

I didn't need to look at the numbers on the houses as I drove down the street because I could hear the party quite clearly. No wonder we'd had complaints. It was almost 02.00 and it was still going strong by the sound of it.

As I walked through the front garden gate I passed a, not so young, couple rapidly propagating the species – rabbits would have had a job to keep up with them – on the front lawn that were completely oblivious to my presence. I made my way down the side of the house, where several guests were congregated and speaking quite loudly to be heard over the music, but no one batted an eyelid. When I reached the rear garden, where the main bulk of party was, it was obvious what the source of the nuisance was. In the garage there was an old Grundig record deck, a fairly new Marshall amp and a couple of – they'd seen much better days, but they still worked – Wharfdale speakers. And the amp had been cranked up to eleven.

When I asked – in a very shouty voice – whose house it was I got the usual drunken bravado, "Who wants to know?"

"I do," I replied.

Eventually I established who the owner of the property was and told them, in no uncertain terms, to turn the music down.

"Why?"

"Because it's 02.00, it's loud and we've had numerous complaints, that's why."

"Who from?"

"Not telling, but on the basis that this is a densely populated residential area I would think you could take your pick." I was really not in the mood.

Some of the guests then started creeping around me asking if I wanted a drink.

"Yeah, why not, I'll join in the party," I announced jovially.

One of them actually went and got me a bottle of beer.

"I was being sarcastic."

They did turn the music off and then announced over the public address system that the miserable bastard from the council (that was me, in case you hadn't realised) had made them stop the party.

To be honest, I was expecting a lot of loud jeering and a bit of hustling as I left, but everyone just kept on drinking as though nothing had happened. They actually seemed relieved that they no longer had to shout to be heard.

When I rang the complainant to let them know what I'd done they said they knew because they'd just heard it announced.

Chapter Twenty-Six

Over the years, more than I care to recall, I've been asked to teach an array of health related subjects in a variety of different settings to a diverse range of age groups. These have included: HIV and AIDS awareness to youths; personal hygiene to infants; healthy eating to all age groups in primary schools; the work of the EHO to secondary school pupils and careers evenings; global warming and the greenhouse effect to junior school children as well as talks to the WI and the business community on the work of the EHO and specifically, food safety and health and safety.

Here's a sample of some of the more memorable talks I gave for your delectation:

I once gave a talk to a reception and year 1 group of children on the importance of personal hygiene. Not a problem I thought – I should have thought a bit longer, and harder, because it did turn out to be a problem.

I took a special liquid – Glo Germ – with me, which is a visual aid for teaching effective hand washing, that glows, like fireflies, under UV light, to see if the children had, or hadn't, washed their hands properly.

The way it works is: the 'guinea pig' – sorry, the volunteer – smears a drop of the liquid all over both hands covering as much of them as possible. They are then placed under an ultra violet (UV) lamp so everyone can see the 'glowing germs' before they are washed away. The volunteer then washes their hands, as they would normally, and then places them under the UV lamp to see if the 'glowing germs' have gone. Inevitably there were some, usually a lot, left behind. If they'd actually been real germs then the potential to spread them about was now obvious.

I've also done this exercise, many times, with adults as well, and they weren't that much better at washing their hands than the children to be honest.

When I asked the children, "Why do you think it's important to wash your hands after going to the toilet?" I immediately got a large show of hands.

That's encouraging, I thought.

"So you don't poo in your pants."

"So you are not sick."

"So you don't wipe poo on your clothes."

Kids are brilliant aren't they? They're not afraid to say exactly what they think. And when you come to think about it, but not too hard, all their answers were in fact correct. Perhaps adults would have phrased them slightly differently though.

I also asked them why they thought it was important to keep their nails short and clean too.

"Because if you scratch your bottom with long nails worms will get under them," was the answer blurted out by a scruffy little girl at the front of the class.

I glanced up at the teachers momentarily who were lined up at the back of the classroom and looking at me as much to say, 'get out of that one'.

I also took disclosing tablets with me to see if the children cleaned their teeth any better than they cleaned their hands. They didn't, but thought it was great fun to have stained teeth and gums.

The object of the HIV and AIDS training was to help youngsters understand exactly what HIV and AIDS was and how they could (but probably more importantly how they couldn't) catch it and to scotch any prejudices they might, and did, have.

At one particular session a geeky looking youth actually thought that everyone who was HIV positive should have it tattooed on their forehead so that everyone else knew. As I explained, not everyone who was HIV positive knew they were, so that wouldn't work, irrespective of the moral and ethical issues.

I used to do a simple exercise with those groups where they had to imagine that a girlfriend (whether they themselves were male or female) was diagnosed as being HIV positive and asked them for their initial reaction. They would all, without exception, immediately assume that she had been sleeping around and had not been practising 'safe sex' and therefore had no sympathy for her. Some even used the words 'slut' and 'slag' to describe her.

I then went on to explain that the reason she'd become HIV positive was because she'd been raped by someone who was carrying the virus. Their attitude changed instantly and everyone was suddenly sympathetic to her plight. This only went to prove that you can't make an informed decision without all the facts.

The most frequent talk I gave was about the work of the EHO. Amongst other groups, this talk was given regularly to students who were considering their options at school for their future careers.

One, rather affected (no not infected), girl asked me, in a rather posh voice, if the CIEH was affiliated to WWF. I couldn't resist. "Why would we be affiliated to the World Wrestling Federation?" I asked whilst trying my utmost to keep a straight face.

She looked at me somewhat puzzled for a moment or two, giggled childishly, and then said, "Ooh you are awful." (I don't think her name was Mandy though, and she didn't say she liked me either). "I mean the World Wildlife Fund."

"Why would you think we're affiliated to them?" I was eager to know.

"Don't you look after the little birds in the hedgerows and stuff like that?"

"No we inspect dead animals in the slaughter house and look down manholes at blocked drains." Harsh I know, but it was the truth.

She wasn't in the least bit impressed. Couldn't understand why!

I don't think environmental health was going to be the career for her. Most of them just wanted to hear the gruesome stories.

Teaching about global warming and specifically the greenhouse effect can be difficult to a group of adults, after all it's not really the sexiest of subjects – you think it is, then you need to get out more; but to a bunch of 10 and 11 year old school children, it wasn't going to be the easiest of tasks that was for sure. But along with Anneka, I was always up for a challenge, and a treasure hunt if anyone is organising one.

Different people learn in different ways. I personally learn more using a combination of verbal, visual and physical learning styles as I find verbal alone can be a bit boring. So because they were my preferred learning styles those were the ones I was going to use with the kids.

So to demonstrate the greenhouse effect effectively I took a king size duvet, 15 tog, and a digital thermometer with me to the school. I asked for two volunteers to go under the duvet and was immediately greeted by a sea of hands in the air, that George Michael would have been proud of ("I want to see every single pair of hands") as every single child – none of them were married – wanted to participate. Once the chosen ones (not by destiny to stop an impending disaster) were suitably ensconced under the duvet I took the temperature outside of it and then under it to show the difference explaining that the duvet represented the layer of pollution in the atmosphere that cloaked the earth and kept the heat in i.e. the greenhouse effect. They understood the concept straight away.

Obviously one of the children farted under the duvet, I could have put money on it, quite loudly and much to the consternation of the other child, which was met with general hilarity. I then went on to explain about methane being one of the greenhouse gasses and where it came from, other than children's bottoms.

Healthy eating was the hot topic at the time. It seems it still is with the rise in obesity and diabetes. However, making the message palatable to a wide range of age groups was a problem I had to chew over for a while.

For nursery children I took a selection of simple drawings of fruit and veg for them to colour in and some soft toys including Sally Swede, Colin Carrot and Bertie Broccoli who I pretended talked to me, when I was with the children, at no other time – honest. They would tell me to tell the children to eat fruit and vegetables because it made them healthy. Seems a bit odd I know, vegetables saying please eat me, but the children really loved meeting them especially when they were allowed a cuddle. At the end of the session there was always something scrummy to eat, and some fruit – only joking.

With the older children I would play games with them including place the food item. I would have the five food groups (fruit and vegetables; bread, other cereals and potatoes; milk and dairy products; foods containing fat and foods containing sugar; meat, fish and alternatives) written on plaques. I would split the class into teams and then give them a selection of foods that they had to place on the correct plaque. The team with the most correct were the winners. Some were very easy like a banana, whilst others were a bit trickier. It's surprising the number of children (and adults actually) that think eggs are dairy products.

Another team game was to give one member of the team the name of a piece of fruit or veg and then they had to describe it to the rest of their team without actually naming it. It was a sort of fruit and veg charades, but with words. I can't imagine playing it with just a mime. Now there's a thought for next New Year's Eve!

Chapter Twenty-Seven

At the turn of the century – the 21st that is – the local health authority successfully applied for government funding to undertake nine different health promotion projects across the county. One of them was a food project which was to focus on the Borntown Council area and had a not inconsiderable amount of, ring fenced, money attached to finance it – although surprisingly, there were only two objectives in it. The first was concerning fruit in primary schools and the second healthy food options in restaurants.

I know, it all sounds very serious and grown up and for that I can only apologise, but never fear, normal service will be resumed as soon as possible.

Initially the economic development department of the council were contacted and told that they would be given the money to progress the project. The Head of Economic Development was a little confused by this news – to say the least – and contacted me to ask what I knew about it. I told him; "I know nothing," (in a very unconvincing Spanish accent) as I'd not been involved in the bid. He, quite rightly, pointed out that the project would sit better within the remit of the environmental health department and that I was now faced with two choices; (i) I could say thanks, but no thanks, or (ii) I could take the money and run.

Well, what would you have done under the circumstances?

I obviously took the money, it would have been rude not to, but I didn't run. I had nowhere to run to, or the right shoes to run in either. And let's face it; it's not every day you get given £20,000 to spend over a three year period without even lifting a finger – or any other part of the anatomy for that matter.

The specific targets were: (i) 50% of primary school children should be eating one piece of fruit twice a week.

Not very ambitious!

Surely that should have been 100% of primary school children should be eating five pieces of fruit (and/or veg) every day.

I believed in aiming high – ever the optimist – and (ii) more restaurants should be offering healthy food options.

Unfortunately the term 'more' was not defined which I thought was a bit stupid really because 2 would have been a 100% improvement over 1, but not really that impressive numerically. Statistics eh!

The first problem I encountered, immediately after agreeing to take on the project, was that I had £10,000 to spend in the first year and we were already over half way through it. I somehow, using my legendary wily ways, managed to persuade the 'powers that be' at health authority to give me – not me personally mind you – the money in the form of a grant. This was so I could carry over any surplus funds at the end of the financial year, unlike surplus council monies which were confiscated – there was never any incentive to save money, only to waste it!

I also decided to add a project of our own about healthy eating in general or 'Healthy Eating on a Budget' as I decided to call it, (catchy don't you think – just like a cold) to the list of things to do. Well, why not?

By the end of the second year I still hadn't spent all of the £10,000 from the first year. Some would say it was because I was a tightarse – to use the Australian slang. I – on the other hand – would have gone with being frugal, whereas my mum would have said that I was just being a careful Christian (even though my name wasn't Christian – other belief systems are available). One thing was for certain, I was not going to waste the money on any unnecessary expenses.

It had been suggested, in the original bid, that some of the money could be used to employ a part time administrator. What on earth for? I would be able to manage what little administration there was myself. I just wanted to spend the health authority money on promoting healthy eating and the council could match fund it by paying for my time – which they were already doing, so there was in fact no extra cost to them anyway.

At the end of the second year I received another 'untouchable' grant of £6,000 to add to the money I hadn't spent from the first. By the end of the third year though, the health authority asked if they could keep the final instalment of £4,000 for another project as I was still so much in credit. How could I refuse? After all, it wasn't my money.

When I retired, some 11 years later, there was still almost £5,000 left in the kitty – which was the 'purrfect' place to keep the money having moved it from under the mattress – even though I'd funded so many, varied, 'food' activities. So I gave the 'leftovers' to a number of the organisations I'd worked with, in the area, during those halcyon days of my career.

My one and only grumble about the whole thing was that no one from the health authority monitored what I was spending the money on even though I kept every single receipt for every single purchase I made. I naturally kept them up to date – once a year – with what I'd been up to, but to be honest they didn't seem overly interested. Shocking really when you consider it was public money I was spending.

To launch the much anticipated (well by me at any rate), 'Healthy Eating Campaign' it was decided it would be a good idea to host our very own version of the popular TV show *Ready Steady Cook*. However, to prevent any copyright infringements we (and that's the royal one) decided to call our event 'Get Ready, Get Set, Cook.'

I spoke with the Head of Community Development; Kate Speed, to see if she could suggest where we could, but more importantly where we should, host the event to reach our target audience – families on low incomes with poor diets. Without hesitation, she suggested a community centre on a small estate, made up predominately of council houses, in the middle of an affluent ward. (A ward is what a council area is split up into for electoral purposes). The estate did not score well when using 'The English Indices of Deprivation' although the score for the ward as a whole was good. Because of this anomaly the estate was regularly overlooked for such events in favour of badly scoring wards.

A working party was set up (it was local government for goodness sake, working parties were compulsory – first meeting; establish a name for the working party) and the inaugural meeting was held at the aforementioned community centre.

There were representatives from the local College of Further Education (who were going to provide the two chefs from their catering department), the local football club (who were going to provide the celebrities in the guise of two of their footballers), the local NHS community dietician, Yvonne (who would give us the necessary dietary advice to make sure that what we were planning was indeed healthy – you can imagine the publicity if it hadn't been) and of course the local authority environmental health and community development departments.

Although Yvonne had lived on the south coast for more years than she'd lived in Scotland, she hadn't lost her Scottish accent. For some unknown reason Scots never do; Sean Connery in *Hunt for Red October* being a prime example. One of the great things about her – there was more than one – was her talent for explaining complicated nutritional information in a simple and easy to 'digest' way. You might say that 'she had us eating out of her hand' – I certainly wouldn't because it would be far too twee.

The one thing (only one) that did seem strange though was now I had a considerable pot of money to purchase services and products with, no one wanted paying. Everyone, without exception, was donating their time, resources and expertise free of charge. Typical, you could have bet your bottom dollar that if I hadn't have had the dosh nobody would have wanted to know.

The College were also going to provide video cameras and TV screens, from their media department, so that the audience could actually watch the food being prepared from above.

It was all very professional – as you can imagine.

No?

Then you; 'must try harder' and 'see me later'.

They were also kindly providing us with a 'Hostess' (no not that kind) or compare in the shape of Ruth Grove, who had plenty of shape believe me, and turned out to be someone I knew from an Advanced Food Hygiene Course I'd taught a few years earlier – yes it really is a small world.

Ruth was two thirds of a femme fatale (a woman who had the intelligence and sex appeal bit, but not someone who lured men to their death – not that I was aware of anyway). She had legs that went all the way up to her derrière; a derriere that needed a step ladder to reach. Okay so I'm exaggerating – a bit – but you get my drift. She was slender and had long dark wavy hair that cascaded down her back like an ebony waterfall. She also had blemish free olive skin and wore black eye liner (or Kohl as it's called in the Middle East) which all went to make her look like Cleopatra – the last pharaoh of Ancient Egypt and well know femme fatal! (Just a coincidence – and you can take that as either a question or a statement.) Well, the Hollywood version of Cleopatra at any rate. I don't know if Ruth looked exactly like the real one because I'd never actually met Cleo (as her friends used to call her). Look I know I'm old, but I'm not that old.

One of the local supermarkets (Tesco – why not give them a plug?) was going to provide all the food for the launch for free (every little helped), even though I had the money to pay for it. So it truly was a community affair.

I was never entirely sure if the TV chefs knew what ingredients their celebrity guests were going to bring along with them – although I did have my suspicions. Can you imagine how embarrassing it would have been if they hadn't had any idea what to cook with the ingredients they were presented with? Our chefs certainly knew what food they were getting, because they'd actually asked for it. The reason being was; they had to agree with the dietician that what they were going to cook was in indeed healthy or the whole event would have been a complete sham – and we couldn't have that. They also needed to know that they would have enough time to cook their dishes in the quantities we required, because we wanted the audience to be able to taste the food too. And on top of all that; they would only have two single ring 'Campingaz Camp Bistro' (you can make your own joke up here) cookers each to cook the food on thereby demonstrating that you didn't need some highfalutin, all singing all dancing, range and a host of other expensive gadgets to produce a well-balanced, cheap, healthy meal.

On the day of the event I was really nervous even though I wasn't actually performing myself. After so much time and effort organising the bloomin' thing I just wanted it to be a success. You might say that I was suffering from the symptoms of food poisoning even though I didn't actually have it. Comprende?

You didn't know that I was bilateral did you?

All the preparations were complete. There was nothing else I could do except sit back, watch and enjoy. It was now in the lap of the Gods.

When I said earlier that both chefs were coming from The College, I was in fact wrong and it really pains me to admit that. Mark Heaney was indeed a chef at The College (senior lecturer in fact), but Anthony Hale wasn't, he was Mark's mate and the Executive Head Chef at the only five star hotel in the town – very posh, very upmarket.

Mark was a rather stout (but definitely not bitter) chap to say the least. He obviously tasted everything he cooked and not just a soupcon by the look of him. He probably tasted everything everyone else cooked as well; it certainly wouldn't have surprised me. The buttons on his brilliant white tunic were fighting hard to keep it fastened, and were only just about winning. I reckoned that one more teaspoon of jus, or wafer thin mint even, and it would have been goodnight Vienna. And I'm not talking about an evening farewell to Rigsby's cat either.

Anthony was the complete antithesis of Mark. He was incredibly slim and his tunic fit him so perfectly that you would have thought it was made to measure. This was probably because he had some underling to do all the tasting for him. Alternatively he was bulimic.

One of the vegetables that Mark requested for his dish was a mooli or long white radish as it is also known. It actually looked like a giant dildo – whatever that is. Needless to say it caused much hilarity and he really cooked up a storm with it. Funny name for a dish I know, but there you have it.

So we had Cleopatra – well her look alike anyway – and Mark Antony, or should I say Mark and Anthony (which sounds like a hairdressers – just a trim for me please and something for the weekend), what could possibly go wrong?

In the end I needn't have worried about anything because the event was an unmitigated success – way, way beyond my wildest dreams (and I have some pretty wild dreams I can tell you).

But why wouldn't it have been? I was working with the best bunch of professionals, who were the loveliest of people as well, on the planet – well the universe to be honest. And that is not an exaggeration. Both chefs cooked fantastic 'healthy' food and were thoroughly entertaining to boot (I must stress at this point that no boots were harmed during the event). The footballers turned up at the right place and at the right time – which was a bonus – even if they did forget their lines when asked how much they'd spent at the supermarket. But it didn't matter, everyone had a great time and Ruth really was THE 'Hostess with the Mostess'. I couldn't have wished for more. Well perhaps Nigella covered in…let's not go there.

When it came to the voting we used cards that had the football team's home and away strips on to choose the winner and the result was pretty close I can tell you – even if it did look more like a rugby score.

'There's no such thing as a free lunch,' according to the old adage, unless of course it is a free lunch. And that's exactly what I gave the participants of the 'Healthy Eating on a Budget' cookery sessions at the local Sure Start centre.

Sure Start centres were created in 1998 to provide a wealth of services and support for families with children under the age of five. These services were to be based on the needs of the local community and their aim was to give children the best possible start in life.

Following the successful launch of the 'Healthy Eating Project', I was asked by the manager of the local Sure Start, Bryan Andrew, if I would provide some cookery sessions for the mums (later on we did one for the dads as well – but in the evenings) who used the centre. Some of them were single, others had partners, but the one thing they all had in common was that their income was limited and so showing them how to provide healthy meals that didn't cost the earth – I didn't go into food miles – was the priority.

The kitchen at Sure Start had been designed as a café kitchen and most of the time it was run as that. It had two cookers, one was all electric – I don't particularly like electric hobs – and the other had a gas hob with an electric oven. There were also two microwave ovens, a large fridge, a large upright freezer and an industrial dishwasher.

There was enough space in the kitchen to accommodate a group of eight people, plus me and Laura, who ran the café; she did a lot of the tidying up after us – what a star.

I worked closely with Yvonne again – mainly because her office was so small – and even persuaded her to make a guest appearance and have a coffee with us at one of the sessions with each group. This was so the 'cooks' could ask her any nutritional questions that I was incapable of answering. It's always better to get it straight from the horse's mouth don't you think? Not that I'm suggesting that Yvonne had any equine characteristics whatsoever – nay, not at all.

I have to point out at this point, even though it's rude to point and grossly bad-mannered to do it twice, that I am not a trained chef. In fact I have no formal qualification/training in cooking, or baking, at all. My one and only teacher, and mentor, was my mum who was a cook both at work and at home – and an absolutely fantastic one too (three four).

'When I were a lad – and I'm not even from Yorkshire – I used to dream of cookery lessons at school.' However, boys weren't allowed to take domestic science, as it was called back then, or needlework as an option. It had to be either woodwork or metalwork and vice versa for girls – very sexist. But needless to say, it's been really useful because I've lost count of the number of armoires, credenzas and wrought iron gates I've made since I left school. Mathematics wasn't one of my better subjects either!

I was, and still am, just an enthusiastic amateur who enjoys the smell of a greasy plate and the roar of the crowd.

I was conscious – which is always handy – that the sessions had to be about what the mums wanted to make, within reason, whilst introducing them to new ingredients, simple recipes and quick and easy techniques; all the time being aware of the cost. But my initial 'food for thought' was how I was going to start it all off when I'd not even met any of them beforehand.

After I'd deliberated and cogitated (I am a gross man) I decided that the first week I would make a homemade soup to take along with me. I would liquidise it then see if they could guess what ingredients lurked within it. At least I would get them tasting and talking about food from the very start. To make it a bit more fun I made a vegetable soup using as many seasonal root vegetables as I could get hold of. I didn't use any leafy vegetables like cabbage, but you could do if you wanted to.

The evening before the first session I was busy peeling and chopping my selection of vegetables in my kitchen at home. A very therapeutic pastime I find. It can also help alleviate the frustrations of the day if you think of someone who's annoyed you whilst chopping the tops off the vegetables. (Is that known as a Henry VIII complex? Or is it a syndrome?).

Now I know you're curious (some would say nosy, but we'll just ignore them) what vegetables I used and how I made it. So out of the goodness of my heart (which sounds like a type of stock) I'll tell you at the end of the book.

The following day I met the eight ladies – but not in the Little Britain sense – at the centre who were going to join me for the sessions. I heated the soup up until it was piping hot having reached a core temperature of 70 degrees centigrade for at least two minutes – not only do as I say, but also do as I do. I served it with some nice (why would I have served horrible) crusty bread – which was shop bought and not homemade.

They all loved the soup and didn't realise how filling it could be. There was plenty to go round and still some left over, which was fortunate because they all wanted seconds. Greedy lot!

The following week I was going to demonstrate how to play the recorder by blowing into the mouthpiece and opening and closing the holes with my fingers.

When it came to them guessing which vegetables were in it, it was really quite amusing. At first they didn't have a clue. Then after they'd thought about it for a few minutes they started to name the more obvious ones like onions, potatoes and carrots. In the end they were just naming any vegetable they could think of, some of which were in the soup and others that weren't. So it was more a case of luck than good judgement when they got one right.

One lady on hearing there was celery in the soup declared that she didn't like celery and certainly didn't eat it (I rest my case – 'cos it's really heavy) and then proceeded to laugh hysterically. This wasn't as a result of eating the celery, but because she'd realised she'd eaten it – I think. But then again perhaps the former was the reason she didn't eat it in the first place.

After we'd finished stuffing our faces I asked them what they'd like to make the following five weeks. They all looked at me completely bewildered. They hadn't realised that that was how it was going to be. They'd assumed it would be like school and I would be telling them what we would be doing.

"Pleeeaaasssee!" I implored them, "We're all adults now and not at school."

They quickly, and democratically, decided that the following week they would like to make that stalwart dish and perennial favourite that should be in everyone's repertoire: spaghetti bolognese. And who was I to deny them?

"Okay," I said, "But what I'll do is show you how to increase the vegetable content of it, which will contribute to your five a day, and how to make it go a bit further quite cheaply." I knew that they were intrigued. How? Because they said so - that's how. Following our little chat, and a cup of tea/coffee, I produced a shopping bag (reusable of course) with all the ingredients they needed to make the soup themselves, and that's exactly what they did.

Before I started the sessions I had a few preconceived ideas of what issues I would have to address i.e. explaining the five a day message and what constitutes a portion of fruit or veg (a handful – the bigger the person the bigger the hand and therefore the bigger the portion). What the 'Eat Well' plate was and how to get a balanced diet etc. But I was wrong; they were all generally au fait with those messages. Whether they were abiding by them was another matter entirely.

The one thing I hadn't considered I would need to address though was basic food preparation and cooking skills – and I do mean basic. Some of them hadn't the slightest idea how to peel or chop vegetables because they'd never been taught how to do it at school during their food technology (as it was called by then) classes. Probably because they were too busy designing food packaging, a skill that would obviously come in very handy on a day to day basis in later life. Also, because their parents hadn't learnt these skills when they were at school, they hadn't been taught them at home either.

That is not a criticism (well the food packaging bit is), just a mere statement of fact.

When I showed them what to do they picked it up quickly and easily and then really enjoyed the whole process of food preparation. But more importantly, there were no missing fingers at the end of the sessions, not even The Slightest Nick – that well known illusionist. Their biggest concern was whether to chop the vegetables into small or large chunks.

"It's up to you," I told them, "Whatever you think is best."

To speed the cooking process up I took my trusty – not the brand name – pressure cooker along with me. Some of the mums had never seen one before, never mind cooked with one. They were all a little apprehensive of it at first, especially when it started to hiss – like Sid – incessantly as the weight on the top started to wobble about erratically. And no, I am not referring to twerking. But they were all amazed at the speed with which it cooked. I explained the nutritional benefits of steaming food too.

I'd also taken some small plastic containers with me so they could take their soup home with them for their families to taste as well. Needless to say I was careful to let it cool down first so that they didn't scald themselves.

The feeling of pride at what they'd achieved at the end of that session was palpable. And they'd only made soup. How were they going to feel by the end of the next five sessions?

The following week when they all returned, and they did all return, they each told their own story of how the soup had gone down at home – and not come back up!

Everyone, without exception, said their families had loved it and that very proud of their mums/partners for making it.

Moments like that just made it all worthwhile.

I have to make it quite clear that I have no intention of turning this into a cookery book, although if I did happen to write one I think my nom de plume would be something like Marella Fearnsey Slaver.

I suppose I could write two books, one with recipes using a pressure cooker entitled: 'Cooking Under Pressure' and the other one not using the pressure cooker entitled; 'No Pressure to Cook.'

So the following week it was Spaghetti Bolognese.

I was going to show them how to make one sauce that could be used three different ways: (i) basic sauce on spaghetti – Spaghetti Bolognese (ii) basic sauce layered between lasagne with a cheese sauce on top – Lasagne (iii) basic sauce with kidney beans and chilli added served on a bed of rice – Chilli Con Carne.

Again the recipes are at the end of the book.

I took my pressure cooker with me again to speed things up and the skills they'd learned the previous week were put to good use again that week.

They made all three dishes and then sat down to eat the fruits – literally in the case of the peppers, courgette, aubergine and tomatoes – of their labours and it was all totally scrummy and again they were knocked out by what they'd achieved.

It then emerged that a concern, of theirs, was the amount of food that they thought was being wasted. When we'd chopped the selection of vegetables, for the sauce, there were inevitably some left over's. One mum said that she thought it was all far too expensive and when I asked her why, she referred to the leftover food. "That will all go to waste," she chided.

"No it won't," I tried to assure her, but she looked sceptical.

I hate waste, not just from the monetary point of view, but because of the 1.3 billion tonnes of food a year that is lost or wasted due to overbuying and the discarding of out of date/rotten food. That's enough food to feed the world's 842 million hungry people four times over (accurate at the time of writing) – and I certainly wasn't going to be contributing to it.

"The left overs could be used to make a ratatouille or the topping for a pizza," I explained. It was then that I realised that usually if they didn't use all the ingredients for a recipe they would simply just throw the excess away. They didn't plan meals for the week and how to use all the ingredients for other things, so I suggested that the following week we could make dishes with the 'left over's', which they thought was a great idea. During that conversation another mum told us how she used over ripe fruit to make smoothies or milkshakes for her children, which was yet another great idea.

The following week they all came back again and told me how they'd made the spaghetti bolognese for their families and how it had been another rip-roaring success.

I very rarely use recipes and if I do, I inevitably only follow them to the letter the first time and then change them by missing out ingredients or substituting them – baking being the exception. I was mindful, however, that the group would probably be using recipes at home. I told them how they could pick them up for free from the supermarkets and also find them in their weekly magazines or on line. I, therefore, took a recipe with me in the latter weeks, but deliberately missed out some of the ingredients (1 teaspoon of balsamic vinegar – expensive if you never use it again) or substituted them, veggie mince being the most obvious example. I wanted to give them the confidence to do the same.

One week I even managed to persuade a mum, who was from a different ethnic background to the others, to cook one of her native dishes for us, and even though she was nervous about it, she did admirably and everyone ate it with Gusto – it was nice of him to drop in.

Over the remaining weeks I did less and less supervision with them until the very last week when I just gave them a recipe and the ingredients and told them to get on with it, which, to their credit, they did. I just stood and watched eight women, from the café side of the serving hatch, whose cooking confidence had gone through the roof, prepare a delicious, healthy, cheap, home cooked meal that they were all, quite rightly, proud of – as was I.

Each week I would also take some, usually new to them, exotic food (more often than not fruit) in with me for them to taste Pitaya or dragon fruit and star fruit being just two examples. Most of the foods I took were quite expensive and certainly too expensive to buy and then realise you didn't like and then have to throw away – I've already done the wasted food bit. This way, if they did like the food then they might treat themselves, and their families, to it at a later date.

By the time we'd finished the six sessions, everyone wanted a pressure cooker (wish I'd had shares in the company that made them) as they could see their nutritional, fuel and time saving benefits – and a few of them actually invested in one. I even demonstrated how to make a complete meal – other than a stew – in it. I steamed chicken breasts with potatoes and carrots all in six minutes – and the chicken was as moist and tender as you like, but most importantly it was thoroughly cooked. I only needed another pan to make the gravy in.

After the first set of sessions, the word got out how much fun they were and all the subsequent ones 'sold out' as soon as they were advertised. Not that anyone had to pay for them of course, I funded them from my large wad (behave) of money.

But it does sound very 'rock n roll' don't you think? Although I have to say I never actually started any of them by shouting; 'Hello Sure Start'.

As part of their rehabilitation back into society, and how to look after themselves, I was asked to provide some cookery sessions – similar to those at Sure Start – for ex-convicts who had recently left prison and were now out on probation.

The Probation Service didn't have their own kitchen so they paid to hire the facilities at the local YMCA – where I believe it's fun to stay. It was only the equivalent of a small domestic kitchen – think more Barratt homes than Buckingham Palace – which was absolutely fine, but it did limit the numbers I could have in there at any one time.

The first group of four, men, that turned up did, and I know you shouldn't judge a book by its cover, look like your stereotypical thugs – think of Britain's most notorious prisoner: Charles Bronson. They had short cropped hair and wore; shrink-to-fit denim jeans with clip on braces; stone washed denim jackets over granddad shirts and Doc Martin (bovver) boots. They had all obviously 'shopped' at the same exclusive boutique. One even had the customary 'Love' and 'Hate' gouged into the fingers of each hand. I don't think they were professional tattoos. If they were, then I'd have demanded my money back. He wasn't exactly like 'Vyvyan' from 'The Young Ones' – no star studs in his forehead or orange spiky hair for starters – but he was as near as dammit.

The biggest of them – 6' 4'' in his stocking feet, approx, and not a gram under 17 stone – opened the proceedings by announcing, threateningly, as he met me; "I don't do cookin'." He was like the local unneutered tom cat marking out his territory.

"That's alright," I replied, unthreateningly; "I don't do washing up."

I know, brave or what!

I started the first week of this six week 'stretch' in exactly the same way I did at Sure Start and amazingly got exactly the same results. The difference was in this case their knife skills were considerably more honed than those of the ladies at Sure Start. I wasn't sure if that was because they had worked in the kitchens of their respective institutions or if it was for some other reason – and I didn't really want to know either.

Again, whilst they were making the soup I asked them to think about what they would like to cook the following weeks.

One bright spark suggested steak and chips. *Dream on.* But he wasn't surprised when I said, "No."

Another bloke pulled me, discreetly, to one side and asked me, very politely, "Can you show us how to make macaroni cheese next week please?"

"Is there any particular reason?" I enquired being naturally inquisitive – okay nosy.

He confided in me that he didn't get to see his children very often and that they certainly hardly ever stayed with him. The weekend after the next session they would be staying with him and macaroni cheese was their all-time favourite food. He explained that he really wanted to be able to make it for them instead of having to provide it out of a can like he usually did.

How could I refuse such a heartfelt and genuine request?

"Of course we can," I assured him.

After discussing it with the rest of the group they agreed that macaroni cheese would be one of the dishes they would make the following week. But because that wouldn't take very long I told them they could make something else as well. I'm generous like that.

We did make the MacCheese. (Now you have a choice here: it can be either a Scottish version of the coagulated milk protein casein pasta dish or alternatively a well know burger chain version – it's up to you). The week after when the bloke came back he told me that he'd made it for his kids and they thought it was the best they'd ever had and he was the best dad in the entire world.

I must admit I did get choked up and had to cough quite vigorously to get rid of the lump in my throat. You just can't buy moments like that. It's what made it all worthwhile.

A week later another one of the guys came up to me and said, "I've started to cook for myself at home now and it's much cheaper than buying takeaway food and tastes so much better too."

My work here is done!

At the end of each session I left them all to wash up – because I didn't – and the kitchen was always considerably cleaner when we left than when we arrived. They would always wash the floor every time as well even though they hadn't split anything on it. When I asked them why they were doing it, I was reliably informed, "That's Borstal training for you."

Perhaps some food business owners should 'do time'. At least they'd learn how to clean properly!

During our tea break, while the food was actually cooking, they would tell me stories – never to be featured on Jackanory – of why they'd been 'put away' to use their terminology. It was generally too much information; believe me, far too much information.

At the end of the six weeks everyone, without exception, thanked me for taking the time to help them and not writing them off as a 'loser'. Considering their visibly hard exterior they were all really quite soft inside, a bit like a strawberry cream in a box of chocolates – the ones with a picture of a thatched cottage on the lid. Not that I would have dreamed of saying that to them.

After the success of the first set of sessions I was asked to provide even more for the Probation Service. In total I did ten, and all the clients, some of whom were women, were always absolutely lovely. The only issue I ever had was with one bloke who couldn't get his methadone prescription before he came to the session because the pharmacy had run out of it. As you might expect, he got a little bit agitated – to put it mildly – but not in the least bit aggressive. In the end we rang around the local pharmacies to see who had some in stock and then sent him on his way to get it. The following week he was back to his normal self.

I sincerely hope that they are all still 'going straight' and leading happy, healthy lives.

Word seemed to get around quite quickly, in a good way, about the 'Healthy Eating on a Budget' sessions and as a result I was asked by another organisation, Help the Aged (now Age UK), to provide some for their client group.

When I asked their manager how they'd found out about the project they actually said, and I quote – verbatim, "I heard it through the grapevine." No, their name wasn't Marvin Gaye.

Help the Aged, just like the probation service, didn't have their own kitchen either and so I asked the ever helpful Bryan Andrew, at Sure Start, if I could use theirs – which he readily agreed to and offered it free of charge. What a very nice man.

Having agreed to the sessions I then became concerned about what I could possibly teach a group of senior citizens about cooking. Surely they would be able to teach me a thing or two – or seven or thirty-six.

I was told that the ladies and gentlemen who would be attending the sessions all lived on their own and rarely cooked for themselves. More often than not they would opt for a supermarket ready meal as their main meal, if they had a hot meal at all, and some would just graze on snack foods throughout the day. The men would be widowers, who had probably never cooked for themselves before, so I would have to start from scratch with them. So at least I knew what my 'rules of engagement' were – which was the closest reference to the SAS I'll ever have.

Volunteers would collect the septuagenarians from their homes, although I did discover that a couple were sexagenarians (I know what you're thinking and you can just stop it right now), drive them to Sure Start and then take them home at the end of each session, so fortunately transport to the venue was not a problem.

I visited Yvonne, again – I think people were starting to talk – and asked her if there were any specific nutritional messages I needed to be giving this particular age group. She informed me that generally the elderly didn't take on enough calories each day and so I didn't always have to use low fat foods. She also said that at their age, unless they had any specific health issues i.e. diabetes/allergies; nothing was really 'off the menu'. She also encouraged me to encourage them to drink more as the fluid intake of the elderly was usually well below what it should be, because they didn't like to keep getting up to get to the toilet – especially in the night.

After we'd made the 'customary' soup during week one, which of course they loved, they asked me if they could make a stir-fry one week.

"Of course you can," I replied. None of the other groups had asked to do this before and I thought that we could have a lot of fun stir frying.

I took my wok in the following week along with a variety of ingredients which included; a selection of raw meats, fish and shellfish; vegetables; spices; as well as stir in sauces – soy and teriyaki.

They absolutely loved playing with the foods – even though you shouldn't – and creating their very own unique concoctions. They each made at least three different dishes and I must admit that I was surprised at their use of some of the spices, which they'd never tried before, but were using with Gay Abandon. Unfortunately Gay could only turn up for that session – shame really.

Another week one of the ladies asked if they would be able to make chocolate éclairs.

"I'm quite sure you'll be able to," I replied. "I'm just not sure I'll be able to," I joked – although it wasn't actually a joke. I'd never made choux pastry before in my life.

'But there's a first time for everything,' according to Melissa de la Cruz - apparently. And this was going to be a first for me.

I spent most evenings, the week before we were going to make them, practising like mad at home. It was like preparing for the *Great British Bake Off* – even if it hadn't been thought of then. I just hoped I didn't end up with a soggy bottom – through nerves or under baking.

When we actually came to make the pastry at the session it turned out to be more like 'shoe' pastry and would have been ideal for soling an item of outdoor footwear with rather than filling with cream. I hadn't the slightest idea what went wrong, but we did have a good laugh about it.

Don't worry, the food didn't go to waste, I fed it to the birds – poor things. Please don't report me to the RSPB.

After the 'Silver Service' sessions one of the more progressive local primary schools in the area contacted me and asked if I would 'like' to provide them with some of the, now acclaimed, 'Healthy Eating on a Budget' sessions for their year 5 pupils. How polite and how terribly British of them. But it wasn't really a case of whether I would 'like' to do it or not – I had to do it – after all, it was my job.

The sessions were to be held as an after school club in the 'school dinners' kitchen. I obviously said that I would 'like' to provide the sessions very much and relished the challenge of working with yet another different client/age group.

The kitchen was only used to re-generate school meals that had been cooked at a central hub, but it did have all the facilities that I would need.

The two dinner ladies, who actually ran the kitchen and a parent helper, would be on hand to help supervise the children, for which I was extremely grateful.

All the previous sessions I've told you about lasted for two hours, but these sessions would only last for one. Not a problem, it would just limit what we could make in the time.

Week one was obviously my tried and tested soup week. After they'd made it for themselves I asked the children what they would like to make the following weeks. As you can imagine some, if not all, of the original suggestions were never going to happen. However, during the following weeks, they did make a range of dishes which included pizza on a variety of different bases such as pita bread and chapattis. The children absolutely loved creating their very own toppings from the selection of 'healthy' ingredients I'd supplied.

I would also take in a selection of new foods, to most of them anyway, to have a taste of too. When the parents came to pick their children up each week they would always ask me how I'd managed get their child to eat certain foods – foods that they themselves couldn't get them to eat at home. I never had the answer, but was glad that I'd made a difference to their eating habits – in the short term at least.

Primary school children are quite easy to engage with because they are usually up for anything, but youths are not always that easy. That is unless you're talking about computer games, or the opposite sex. And it's especially difficult when you're at least double their age.

Okay, I know that that's a generalisation – as opposed to General Eisenhower – but a lot of them can appear to have more than a passing resemblance to Kevin and Perry. I know, "That is so unfair."

I was convinced that teaching such a group how to cook healthy meals would not exactly be on the top of their agenda because it was 'so gay' or 'lame' or whatever the hip phrases were at the time. But I was wrong. They so wanted to learn about other styles of cooking including; Indian, Mexican and of course Italian. They had great fun developing their own dishes by combining recipes and creating such delights as 'chilli lasagne' and 'curried bolognaise'. Perhaps that was the start of what is now referred to as 'fusion cuisine'. If only we'd patented the idea.

There was one organisation, which shall remain nameless, that contacted me and asked if I could provide some of the aforementioned sessions for their client group. I naturally said that I would be more than happy to, but would need to see their facilities first to make sure that they were in fact suitable. Also I would need to establish how many of their clients could be in the kitchen safely, at any one time.

The next day I went to their premises and met with the manager. The kitchen was indeed suitable, but, because of its size could only accommodate five people plus me. As the conversation went on the manager informed me that there would be a charge for using the kitchen. This took me back somewhat.

"But I'm providing all the food for free as well as my time," I pointed out in case for some reason they hadn't realised. I thought I'd made it perfectly clear during our initial conversation on the phone.

"We will still have to charge you for the use of the kitchen," was their curt reply.

"The object of the exercise is not for you to make a profit out of the sessions." The response was spat from my mouth as if it had a bad taste to it, which in all fairness it did. But the manager wasn't in the least bit fazed by my barbed reply.

I eventually left the premises, having agreed to disagree with the manager on the subject of charging for the use of the kitchen, and returned to the office. By the time I got back I was starting to think that perhaps it was me who was being unreasonable.

I contacted a variety of people that had been involved with these sessions from the beginning and asked for their opinion on the matter. Every single one of them said that it was completely outrageous to be charging for the use of the kitchen and that I should give them a wide berth. I did finally refuse to do the sessions there, but felt really sorry for the clients who had missed out due to the greed of the manager. And I do believe it was the manager's fault and not the organisations per se.

The manager then went on to accuse me of being prejudiced against their client group and whenever I met them at any health related meetings after that there was always some snidey dig aimed in my general direction.

But it was water off a duck's back – especially when I knew I was in the right.

Now if you recall I mentioned another target that I have omitted to tell you about, up until now, concerning Fruit in Schools or FIS as hip people say, not that I know any hip people.

The FIS project was piloted at a large primary school in one of the more deprived areas of the town and comprised of working with a single year group – from reception to year six – every day for a week. So in that case the project lasted for seven weeks. Each day of their particular week all the children in that year group would have access to a different piece of fresh fruit or fresh vegetable, as well as fruit juice, dried fruit and tinned fruit (in juice). Depending on the seasonal availability I always tried to provide the more exotic rather than the everyday fresh fruit.

Once again I asked Tesco if they would like to support the pilot (because pilots are so poor) as they had already done with the cookery pilot – even though I had the money to buy the produce – and once again they agreed. I have to admit it was great fun cramming a shopping trolley full with around ninety pounds' worth of food at a time and not having to pay for any of it.

I issued the school with a copy of the 'Give me 5 Activity Book' as well. The book was compiled by The British Dietetic Association, and contained a variety of games, puzzles and other fun things to do, for all age groups, that were food related – obviously.

I also gave the teachers ideas of how to incorporate the fruit and veg into the curriculum so that the food wasn't just left there in case anyone wanted to taste it. For the older children I suggested a maths task; finding out the percentage of the class that liked each food (so they all had to taste everything – sneaky). The younger children could look at the geography of where in the world the different foods came from and how they eventually got to the school as well as the biology of how the plants grew. And the very young children could simply draw the fruit and veg before they actually scoffed them and all the time the five a day message should be emphasised.

The local university also came on board to evaluate the project. They spoke with the children at the end of each week to find out what foods they liked, or alternatively disliked, as well as what they had learned about the foods.

Some of the comments from the children included; "I thought a pepper would taste like pepper, but it didn't. I like the red ones best and my mum puts them in omelettes now."

"Prune juice is yuk."

Having said that one little boy liked it so much that he went round the class drinking all the leftovers. He was sent home with a note informing his parents that if he was a bit loose the following day he hadn't picked up a bug from school, it was just an over indulgence of prune juice.

"Do Dragons lay dragon fruit?" Bless!

Eventually 82% of the primary schools in the area participated in the FIS project. Not bad eh?

Chapter Twenty-Eight

Even inspectors get inspected, or audited if you're going to be pedantic –
and that is your prerogative – which was one, amongst many other
purposes of the Food Standards Agency (FSA) – allegedly.

Their audit was quite simply an Ofsted (Office for Standards in
Education) inspection, but for Environmental Health department food
safety teams.

The FSA was founded on April 1st 2000 (I think the irony may have
been lost on them) after several high-profile outbreaks, and deaths, from
foodborne illnesses including Bovine Spongiform Encephalopathy (BSE).
They are an independent government department, no comment, and are
responsible for food safety and hygiene across the whole of the UK. They
are supposed to work with business to help them produce safe food and
with local authorities to enforce food safety regulations – again, no
comment.

The food writer Joanna Blythman has been highly critical in the past
of the FSA (and quite rightly so in my opinion – there, I've commented
now) saying: "It was meant to be a watchdog defending consumer
interests but it seems to have had an unduly cosy relationship with food
companies, bio-tech companies and large retailers ever since it was
created."

The FSA audit Environmental Health departments to check that they
are meeting their inspection targets and carrying out those said inspections
in compliance with the relevant legislation and guidance.

'So what are these targets you talk of?'

I knew you would want to know.

When food premises are inspected they are scored – the lower the
score the better – based on seven specific criteria which include;
temperature control; management systems; structure; cleanliness and
training of staff, and then rated. The ratings are alphabetical and range
from A – E. An A risk category premises will have scored 92 or above

and should therefore be inspected every six months. A B will have scored between 72 and 91 and should be inspected every year and so on and so forth in six monthly intervals. The FSA insist that all inspections due in a particular year should be achieved in that year – irrespective of staffing levels or other any other constraints placed upon the department.

They don't live in the real world.

The hypocrisy of it all is that the FSA set themselves a target for introducing a nationwide 'Food Hygiene Rating Scheme' for food premises and didn't meet it! Not by a long shot.

I personally think that quantitative targets are meaningless, especially if they don't have a quality element attached to them as well. Borntown Council had a target of answering the phone within three rings, but there wasn't a quality element for the response when it was answered. So arguably, the phone could be answered within three rings, the caller could be told to 'piss off', and the phone put down. Technically the target had been met. Need I say more?

When a food premise has been inspected, and the proprietor informed of any contraventions, a revisit is usually scheduled to make sure they have completed the schedule of works satisfactorily to achieve compliance with the law. The FSA say that revisits cannot be counted in the total number of visits made in a year, only the original inspections. So basically if every premise inspected required one revisit – and many usually require more than one – then only half of the total number of visits made in a year can actually be counted. How ridiculous is that?

When I was a student I was always taught that the primary objective of an environmental health officer's work was to get compliance with the law. It would have been really easy to simply make no revisits at all and therefore achieve the required target and just ignore compliance element, but what message would that send out? That we didn't give a toss what the premises were like because we wouldn't be returning to check if they'd done the work we asked them to do. In fact we might not see them again until the next scheduled inspection which could be 18 months away – unless we received a complaint about them – that's what.

I don't think so. Not on my watch. I was interested in outcomes, not outputs!

Because Borntown had nearly 2,500 food premises and only four members of staff able to inspect them, we were lucky if we managed to inspect just our As & Bs, never mind the rest. So inevitably, we seldom achieved our target.

By 'able to inspect' I mean the FSA insisted that food inspectors – EHOs and technical officers with EHORB registration – had to have at least two years experience in food safety inspection and at least 10 hours per year CPD (continuous professional development) in food related matters.

Before our audit I had to fill out an inordinate amount of paperwork. Who said computers reduced it? I had to send them copies of our current Food Plan and because we were being inspected at the beginning of April, I also had to send them the previous year's one as well. The financial year that is, not calendar. This was at the time the department's teams had been drastically – and ludicrously – reorganised.

As a result, my most qualified and competent food officer had been put into the pollution team and some of my food team were now being managed by the trading standards manager whilst I was managing some of the trading standards officers.

I predicted it wouldn't work – and it didn't. If it ain't broke, don't fix it!

The audit was going to be held over two days which meant that the auditors were coming down the evening before (or earlier maybe – if the weather was nice) and leaving the day after the last day's inspection – so a nice jolly by the seaside.

The evening before it all kicked off my boss went into a complete blind panic. He didn't like to be told that there were things wrong with the department – even though there were, everyone could see – and he certainly would never admit it. Well it was too late now because I'd already told them in the paperwork I'd sent. Honesty is always the best policy as far as I'm concerned.

"Don't let them walk all over you," were his words of wisdom as I was leaving the office. There was no chance he'd have been in early enough to impart them the following morning.

The next day I received a call from reception at 09.00 to say that Mrs Greenwood and Mrs Cross from the FSA had arrived.

Let battle commence.

I skipped across reception to meet them (I just needed to practise the hop and jump and I would be well away).

"Good morning Mrs Cross," I said not looking at either of them in particular because I didn't know who was who.

"It's Miss," came the emphatic and resolute response, in a deep husky – at least forty a day – voice, from the larger of the two women as she shook my hand in greeting.

Of course it was, silly me.

Her grip was like a vice.

MISS Cross was a well built, as in muscular not fat, above average height, middle-aged woman with a short back and sides haircut and an enigmatic stare.

I know it's a bit of a cliché, but she looked like a 1980s Russian shot putter, but without the Lycra – thank goodness.

She was dour, and within the hour I found out she was also sour as well as being built like a tower. Eat your heart out Dr Seuss.

She wore a fawn Rain Mac tied at the waist from which fawn linen trousers hung between her knee and her, fawn, solid flat shoes.

"And I'm Ms Greenwood," squeaked the other one as timidly as The Dormouse at the Mad Hatter's tea party.

Ms Greenwood was the complete opposite of Miss Cross. She looked very prim and proper – and timid – as if she wouldn't say 'boo to a goose', in case the goose said boo back I shouldn't wonder.

Having said that, I didn't say boo to the geese I'd met – people in glass houses.

She wore a pleated plaid skirt, or kilt I suppose you would call it, from the Mac Miserable (Victor Hugo's abandoned first draft – it didn't work set in Scotland) Clan I believe.

(If she'd been a woman of the night would that have made her a Tarte tatin?) On top she was wearing a cream, polyester, pie crust collar blouse under a cream, acrylic, Broderie Anglaise cardigan and on her feet, Jesus Sandals. Not quite Miss Jean Brodie, but definitely a B movie version of her.

From now on Miss Cross and Ms Greenwood will be known as 'Tweedle Dee' and 'Tweedle Dum.' Although it was pretty obvious that neither of them was dumb. I just had to remember not to call them that to their face.

The day they arrived was a beautifully sunny and warm, for the time of year, spring day and so I said, quite amiably, "Because it's so nice I've booked a beach hut for you so you can conduct your audit from there."

They both just stared at me with a look reminiscent of someone sucking on a lemon.

Okay, so I've got the cut of your jib – and I'm not that fond of the way it's been cut I can tell you.

It was Miss Cross who interviewed me. Well I say interviewed, it was more of an interrogation really. She may as well have gone the whole hog and tied me to a leather chair (not even remotely funny) with an angle poise lamp shining firmly in my face. She made the Gestapo look like Oprah Winfrey.

Ms Greenwood just quietly waded through the mountain of files that they had requested before their visit.

You may be surprised to learn that the interview didn't go particularly well – but I doubt it. She kept asking me rhetorical questions in between reminding me what I'd written in my report. I wasn't sure if she really expected me to answer the questions, but when I tried to substantiate any answers I did give, she just blanked me. So in the end I gave up.

"You didn't achieve your target last year."

No kidding. You must be an absolute genius to have worked that out. Einstein has got nothing on you.

"I know, I wrote the report that says that."
"You haven't got enough staff."

Really.

"I know, I wrote the report that says that."

When are you going to tell me something I don't know?

"Why is it that your most qualified food officer is now dealing with environmental protection complaints?"
"I don't know you'll have to ask the boss that question."
"It doesn't make sense."

You think so!

"Why are you managing some of the trading standards officers?"
"I don't know you'll have to ask the boss that question."
As you can see, I was being my usual convivial and helpful self.

"Why are some of your food officers being managed by a trading standards manager?"

I should have put my answers on a tape loop, it would have been easier.

The money that was being wasted on this fiasco could have gone to fund another food officer for a few months and then we might, and I do stress the word might, have achieved our – meaningless/pointless/worthless – target.

After an extended lunch hour – it was nearer two in fact – the 'Two Tweedles' (sounds like a circus act, and there's many a true word spoken – or in this case written – in jest) eventually returned and Miss Cross, quite forcefully, asked to see me in their office.

It was then that one of my colleagues, who hadn't yet met them, suddenly joked, "He's really missed you y'know."

I'm not sure if she was meant to hear it, but she did.

I just wanted to curl up and die. Why won't the ground swallow you up when you so desperately want it to?

And so off I went back into Room 101 – well the room didn't actually have a number, but it felt a bit like Room 101 to me.

Towards the end of the day the 'Tweedles' (they've rebranded already) called me in yet again, this time to ask if they could stay behind and work late. Remembering what my boss had said the day before about them walking all over me, I promptly said, "No, I'm afraid" – I wasn't any longer – "your temporary door passes will only work up until 18.00, so you wouldn't be able to get out of the building after that time." That was in fact true. "If you would like to come in early tomorrow morning I'll be here at 07.30 to let you in." That was also true.

They looked at me completely shocked. I'm not sure if it was because I'd actually refused their request or because of my suggestion of such an early start.

When I came out of the room – feeling rather smug with myself it has to be said – I bumped into the boss. He nervously enquired how it was all going as I sauntered nonchalantly back to my desk.

I gave him a considerably watered down version of events, he'd find out anyway the next day when it was his turn to be interviewed. I then told him what had just happened.

He went totally ba-llist-ic.

"You can't tell the FSA they can't stay late."

I just did.

"What will they think?"

If I'm brutally honest, I don't give a shit.

"But you told me not to let them walk all over me", I protested, but only half-heartedly.

"I didn't mean do that", was his garbled, petrified response.

Then what did you mean?

"I'll go and tell them that I'll stay behind for however long they want to work so I can let them out," he said as he went dashing back to Room 101.

What a pussy!

By the time he'd got there though, they'd already packed up.

Ha ha ha hee hee hee an elephant's nest in a rhubarb tree.

Childish I know, but I was beyond caring.

The next day I was at the car park barrier at 07.30 on the dot, as promised, to let 'Dee' and 'Dum' (I could see them going their separate ways very soon) in, but they didn't arrive until 08.40.

So they were good at meeting their own targets then!

As part of the audit they checked a selection of the officer's files – their selection, not ours – to make sure that they were 'up to standard'. I won't bore you with the details of what 'up to standard' means, but suffice it to say they all were; they couldn't find a single thing wrong with them, and they tried. Believe me, they really tried. But I knew they wouldn't.

They also followed one of the officers around during a food safety inspection, again to make sure that they were 'up to standard'. And of course, they were. I knew that as well.

Their main criticism was that we didn't use our enforcement powers often enough and relied too much on informal action.

That's the informal action that actually achieved compliance because we knew the premises and the owners – they didn't.

We hadn't hit our targets, but we were doing a good job. More than good in fact, but they were never going to say that. I don't think they knew what encouragement and praise were.

So they wanted us to go back to the overzealous enforcement practices of the 1980s for which the profession was highly criticised and told to take more informal action.

Some people never learn.

They eventually sent us a report that told us exactly what I'd told them in the report I'd sent to them.

So that was all good then.

I hasten to add that I am not tarnishing all the FSA auditors with the same brush. I am sure that there are some out there who are the complete opposite of the two I encountered, which would only serve to demonstrate that there was a lack of consistency in FSA audits.

On the positive side though, at least I'd recognised what was wrong.

When I asked what help the FSA would be able to give us, to help us achieve our target, unsurprisingly the question fell on deaf ears. Ignorant more like.

How very convenient.

But I'm not in the slightest bit bitter. In fact I've just licked myself and I taste quite sweet actually – but you'd already guessed that. Or maybe it's the perspiration.

The other positive that came out of it was that I got my best food officer back and managed the whole of the food team again and no trading standards officers.

Life was good again.

Chapter Twenty-Nine

So there we have it, a glimpse into my life, well as an environmental health officer anyway. There are obviously plenty of other stories I could have told you but decided not to because they were a bit dull in comparison. Who wants to hear about premises that were okay? I'm sure if you talk to other EHOs they will have very similar and quite different stories to tell if you are willing to listen. Who knows, there may be a sudden rush of environmental health memoirs after this. But at least I can say: "I was first."

All the stories are true, which goes to prove what Mark Twain said: "Truth is stranger than fiction." Only the names have been changed to protect those who need protecting.

If you have read this and think that environmental health is the career for you then I would advise you, wholeheartedly, to: "Go for it." It has been a wonderful career for me, and for the most part I was extremely happy in it, even if it wasn't my first choice. The only proviso I would attach to that is; with government cutbacks at the moment I'm not sure there are that many jobs still available. Certainly nowhere near as many as there used to be. False economy in the long term in my view because when the shit really hits the fan there won't be enough officers to clean it up.

Chapter Thirty

Vegetable Soup

Ingredients:

Oil for cooking (personally I prefer extra virgin olive oil, but use whatever you prefer or whatever you have to hand)

Onion(s)

Garlic

Carrot(s)

Parsnip(s)

Swede

Turnip(s)

Butternut Squash (which is strictly a fruit)

Sweet Potato

Potato(s)

Celery

Vegetable stock

Mixed Herbs (optional)

If you want to use a selection of frozen vegetables instead of fresh (or as well as) then do so.

I was never into precise measurements when I cooked – and I'm still not – the idea was you could use as little or as much of each ingredient as you liked. Basically my mantra was – and still is – 'chop it up and chuck it in'.

Also if I said use two carrots and you had three, then use all three, don't leave one behind all on its own. Not only that, but vegetables come in different sizes so mine might be bigger than yours (ooh err missus). But as we all know – size doesn't really matter.

Soups are a great way to get rid of leftover vegetables, or those that are starting to look a bit sad, and in the case of the carrot above – lonely.

Method:

Pour a drop of oil into a pan – just enough to coat the bottom of it.

Peel and chop the onion(s) – try not to cry.

Peel and crush (or grate) a couple of cloves of garlic (if you like garlicky food then use more).

Add the onions and garlic to the oil in the pan then turn on the heat (low) and start to fry gently.

If you add the garlic to hot oil it will burn and taste bitter.

To stop the onions from going brown add a pinch of salt. This will release the water from the onions. Cook until soft.

In the meantime peel and chop the rest of the vegetables. If you can't get any of them it doesn't really matter. If you want to add something else (instead of or as well as) then do so – don't be afraid to experiment.

Add the vegetables (fresh or frozen or a mixture of both) to the pan and stir until they are all coated in the oniony/garlicky oil.

Cover the vegetables with a vegetable stock of your choice. Make this up according to the manufacturer's instructions i.e. dissolve 1 cube in 190ml of boiling water or 1 pot in 500ml of boiling water.

If you want to add some herbs (to taste) then go ahead, be a devil. However, it's best to do so at the end to keep their flavour.

You might want to add some chilli powder/flakes to make it spicy or some curry powder/paste to make a curried vegetable soup. The choice is yours.

That's the great thing about cooking your own food you can make it taste exactly how you want it to and not how the manufactures do.

Cook until the vegetables are soft.

If you have a liquidiser then use that to liquidise the soup, if not then mash it with a potato masher or if you prefer, leave it chunky.

If the soup is too thick when liquidised then just add some more liquid (I like to use V8 vegetable juice, but water will do). It's much easier to thin a thick soup than trying to thicken a thin one.

Liquidising soup is a great way of getting children, and adults actually; to eat vegetables that they wouldn't normally – because they can't see them.

Be careful not to use too many strongly-flavoured vegetables, like parsnips, if they don't like those though. They may not see them, but they will certainly be able to taste them.

If you want to add peas to colour it green – because they love peas – then do so and tell them it's pea soup. (Don't tell me you haven't lied to your children before – Father Christmas, the Tooth Fairy – need I go on?).

There are no hard and fast rules to making soup. The only problem is you'll never make it exactly the same way twice.

Spaghetti Bolognese/Lasagne/Chilli Con carne

Basic Ragu

Ingredients:

Oil for cooking (personally I prefer extra virgin olive oil, but use whatever you prefer or whatever is to hand)

Onion(s)

Garlic

Mixed Peppers (red, green, yellow, orange – all or any combination)

Courgette

Aubergine

Mushrooms – nothing fancy like shiitake (not the same as taking the piss), simple closed cup will do

Veggie Mince – brand of your choice, or if you're like me – whatever is on special offer. This is ideal for vegetarians – obviously – cheaper than beef mince, low in fat and is free flow straight from the freezer. I've lost count of the number of people I've cooked this for who say they'd be able to tell the difference and can't. It's all about the flavours you add

Red split lentils

Canned chopped tomatoes – the cheapest will do

Mixed Herbs

As far as quantities are concerned, again use as much or as little of each ingredient as you want. However, I do encourage making large quantities of the sauce if you have a pan large enough and a freezer to store the excess. Great for when you can't be bothered – or just haven't got the time – to cook, knowing that there is some nutritious homemade grub in the freezer and all you have to do is defrost and re-heat it.

Method:

Pour a drop of oil into a pan – just enough to coat the bottom of it.

Peel and chop the onion(s).

Peel and crush (or grate) a couple of cloves of garlic (if you like garlicky food then use more).

Add the onions and garlic to the oil in the pan then turn on the heat (low) and start to fry gently.

If you add the garlic to hot oil it will burn and become bitter.

To stop the onions from going brown add a pinch of salt. This will release the water from the onions.

Cook on a low heat until soft.

(Does any of this sound familiar?)

In the meantime de-seed the peppers and chop them along with the rest of the vegetables. If you can't get all the colours of pepper or don't like one of them it doesn't really matter. If you want to add something else instead of one of the ingredients or as well as then do so – experiment.

Add the vegetables to the pan and stir until they are all coated in the oniony/garlicky oil.

Add the veggie mince and coat with the oniony/garlicky/vegetabley oil (new culinary term).

Add enough cans of chopped tomatoes to cover everything.

If you want to make it all go that bit further then add some red split lentils (rich provider of protein and carbohydrates as well as calcium, iron and B vitamins). These will bulk it out but cook down so you hardly notice them.

Cook until the vegetables are the right consistency for you. You may prefer the peppers soft or alternatively 'al dente' (I knew his brother Billy). It really is up to you and don't let anyone tell you otherwise. The perfect way to cook food is the way you like it.

Add the mixed herbs (to taste) at the end to keep their flavour.

Spaghetti Bolognese:

Serve the sauce on spaghetti (like you hadn't already worked that out for yourself) or any other pasta of your choice. Why not Penne Bolognese? Be radical.

Lasagne:

Layer the basic sauce with the lasagne (am I teaching granny to suck eggs here?)

The first layer is pasta, then a thin layer of sauce. Repeat until the dish is full. The last layer is pasta. Make a cheese sauce to cover the top layer then bake in a moderate oven until the cheese has gone golden brown (not a Farrow and Ball colour). A proper Italian lasagne should cut like a slice of cake.

Moussaka:

Substitute the pasta for layers of char-grilled slices of aubergine. (I didn't make this with the group because: (a) I've only just thought of it and (b) we wouldn't have had enough time anyway).

Chilli Con Carne:

Add a tin of kidney beans (cheapest will do) to the basic sauce and some chilli (powder/flakes – which ever you prefer/have to hand) to taste then serve on a bed of rice.

Quick and easy, lump free, Cheese Sauce.
Ingredients:
1 table spoon of butter or margarine – not low fat, it doesn't work
Plain flour
Milk – skimmed if you prefer, that does work
Mature Cheddar Cheese
English mustard (optional)

Method:
Melt the butter. I usually do this in the microwave, but cover the container so that it doesn't spit everywhere.

Sprinkle some plain flour into the melted butter and whisk until you've got the consistency of thick porridge.

Add a quarter of a pint of milk, whisk then heat in the microwave for 1 minute then whisk again to make sure all the ingredients are combined.

Grate some mature cheddar cheese, quantity to taste, and add to the sauce. The stronger the flavour of the cheese the less you need to use and therefore the lower the fat content of the sauce. If you keep adding more mild cheese it will not make the sauce taste any stronger, but the fat content will increase.

Add half a teaspoon of English mustard if you so desire. You won't taste the mustard, but it will make the sauce cheesier (there's a joke here, but I can't think of it).

Whisk all the ingredients together thoroughly.

Keep heating the sauce in the microwave a minute at a time and whisk at the end of each minute and you will never have a lumpy sauce. If the sauce is too thick just add a drop more milk.

Easy Boiled Rice (not a subsidiary of Easy Jet).
Ingredients:
Basmati rice
Boiling water
Method:

Put one cup of basmati rice in a large Pyrex dish (1 American cup measure will serve two people).

Pour two cups of boiling water over the rice and then put the lid on.

So that's one measure of rice to two measures of boiling water.

If you don't have a Pyrex dish, then use a large bowl and cover it with cling film.

Place in the microwave and cook on maximum for 10 minutes without touching it.

Leave to rest for a minute.

Eleven minutes later – fluffy boiled rice (just like Mata used to make).